Engineered Pores for a Greener Future: Applications of Nanoporous Materials

Caspian

Table of Contents

Chapter 1. Introduction

E. H. Cho, Q. Lyu, and L.-C. Lin, (2019) Computational discovery of nanoporous materials for energy- and environment-related applications, Molecular Simulation, 45:14-15, 1122-1147, DOI: 10.1080/08927022.2019.1626990. Reproduced by permission of Taylor and Francis Online

1.1 Overview

A recent study on worldwide energy consumption reported a consistent increase in fossil fuel consumption.[1,2] A typical untreated flue gas emitted from coal-fired power plants is composed of, by volume, N_2 (70-75%), CO_2 (15-16%), O_2 (3-4%), H_2O (g) (5-7%), SO_2 (800 ppm), NO_x (500 ppm), and others.[3,4] As notoriously known, CO_2 is a major contributor to global warming, comprising 81% of greenhouse gases. This corresponds to 6511 Million metric tons of CO_2.[5] Furthermore, even a small amount of SO_2 and NO_x, amongst the six principal air pollutants defined by National Ambient Air Quality Standards (NAAQS),[6] are hazardous to environment and humankinds health. Despite the environmental issues, consistent energy demand seems non-negotiable for the prosperity of humankinds. A few solutions were suggested in order to relieve the energy and environmental burdens, e.g. by the removal of hazardous gas, the use of energy sources other than coal-fired power plants, and chemical processes that are less energy intensive.

All of these solutions can be partially or fully facilitated with the development of nanoporous materials.[7] Nanoporous materials in both two or three-dimensional forms, such as zeolites, metal organic frameworks (MOFs), zeolitic imidazolate frameworks (ZIFs), covalent organic frameworks (COFs), and carbon-based materials (e.g. nanoporous graphene), have gained great attention for their potential in energy- and environment-related processes either as adsorbent materials or membranes.[2,8–10]

The wide applications of various classes of nanoporous materials for energy- and environment-related processes are schematically summarized in Figure 1.1, which overall can be attributed to their many advantageous properties, such as large surface area and porosity, low heat capacity, and selective adsorption and transport of one molecule over another. Another critical feature of nanoporous materials is their chemical and topological diversity. For instance, ~230 IZA (International Zeolite Association) zeolites with distinct topologies have been reported to date while more than 100,000 hypothetical topologies were predicted;[11–13] diverse topological features exist such as one-, two-, or three-dimensional channel system, straight or zigzag channels, cages divided by windows, and others. Furthermore, different Si:Al ratios as well as their distributions for a given ratio in a specific topology can be possible. When Al is present in the system, a cation needs to be introduced in the framework in order to neutralize the charge imbalance. Various cations such as Na^+, Mg^{2+}, Ca^{2+}, and others can exist in a structure. Similarly, MOFs, composed of metal clusters and organic linkers are also highly tunable. Since the first MOF which was synthesized more than two decades ago,[14] tens of thousands of MOFs have been reported.[14–18] Moreover, more than hundreds of thousands of MOFs can be hypothetically achieved by enumerating various combinations of metal clusters and organic linkers.[16,19]

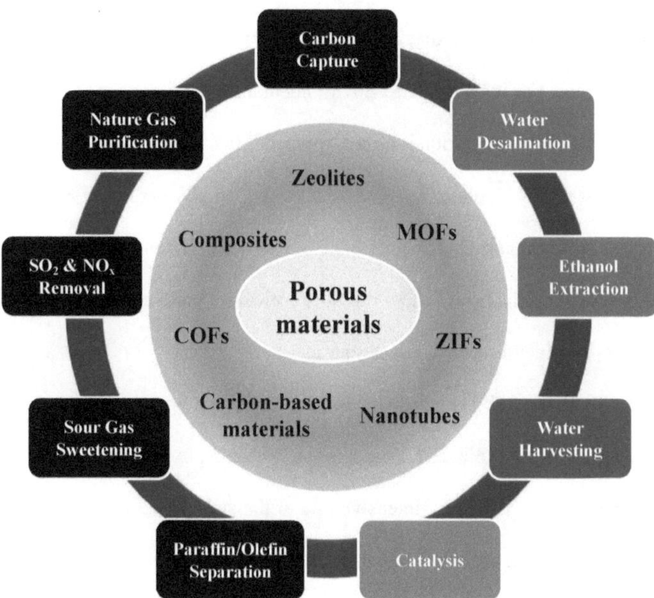

Figure 1.1 Different classes of nanoporous materials (inner circle) and their wide applications (outer circle). This figure does not offer an exhaustive list of possible porous materials and applications.

1.2 Motivation

Natural gas, a principal source of methane (CH_4), can be a great alternative to coals as they generate less carbon dioxide emissions. However, typical raw natural gas can contain hydrogen sulfide (H_2S) at a high concentration, hence is commonly referred to as sour natural gas. The criteria that defines natural gas as sour depends on the country, state, or applications. For instance, U.S. Environmental Protection Agency categorizes natural gas as sour if H_2S concentration is above 4 ppm.[20,21] In contrast, Railroad Commission of Texas defines sour gas pipelines as those with more than 100 ppm of H_2S.[22] Since H_2S is toxic, corrosive, and flammable, the presence of H_2S is deleterious to effective energy generation or chemical productions. Therefore, it is critical to remove H_2S from sour natural gas, the process of which is referred to as sour gas sweetening.

Since 1930, natural gas industry has been relying on amine-based scrubbing for sour gas sweetening, which utilizes alkanolamines to chemically absorb H_2S. However, the amine-based scrubbing is energy-intensive due to the strong chemical interactions between the amine and H_2S, especially when there is a significant amount of H_2S present.[21,23,24] To this end, adsorption-based separation using nanoporous materials has recently drawn considerable attention.[21,25,26]

With the numerous possible structure candidates, a large library of nanoporous materials provides great opportunities to find the best one for a given application. One can explore different classes of materials and/or tune the topological and chemical properties in order to meet the needs that are distinct for different applications. On the other hand, it also imposes significant challenges in the selection of a suitable material by using purely experimental approaches to search over all the materials in the library. As a complementary method, computational modeling methods can largely facilitate the materials discovery and design. One commonly employed tool is quantum mechanical (QM) method, such as wavefunction (WF)-based or density functional theory (DFT) calculations. Although there are some approximations in WF or DFT methods, these calculations can provide accurate results and insights in chemical systems. They are frequently employed for studies including those related to nanoporous materials, such as structure optimization as well as the identification of the most favorable configuration and energy of a molecule adsorbed in nanoporous materials.[27] Despite high accuracy, however, the expensive QM calculations make it prohibited to study systems that have larger dimensions, involve multiple molecules, or require a long-time scale (e.g., adsorption or diffusion of gas molecules in nanoporous materials). On the other hand, classical molecular simulations such as Monte Carlo (MC) or molecular dynamics (MD) are able to efficiently compute thermodynamic and kinetic properties, related to gas removal or storage. In these calculations, interactions between molecules or between the molecule and the framework are described by some potential functions, which are referred to as force field (denoted as FF hereafter). For this reason, classical molecular simulation greatly depends on the accuracy of the FF. With accurate FFs, these state-of-the-art molecular simulation techniques can be efficiently and accurately utilized to compute adsorption and separation properties of gas in nanoporous materials.

Although classical molecular simulation techniques, such as Monte Carlo or molecular dynamics, may be more efficient than purely experimental or quantum mechanical methods, conducting large-scale studies in a trial-and-error or brute-force manner still requires a tremendous amount of computational resources. In light of this, incorporating supervised machine learning algorithms for a significantly more efficient screening study can be a promising direction. Moreover, machine learning methods, such as random forest, can also provide quantitative and qualitative analyses on materials' features that result in useful design rules.[28]

1.3 Structure of book

This book deals with accurate and efficient materials search for various energy and environmental issues, with specific focus on applications to natural gas purification. To this end, this book specifically focuses on two major directions: molecular model development for accurate molecular simulation calculations and data-driven approaches for facilitating materials discovery and design. In summary, this book is organized in the following ways. Chapter 2 specifically introduces a new methodology for molecular model developments and its application to various gas molecules. Chapter 3 focuses on our study of designing descriptors of nanoporous materials and their usage in machine learning for accelerated predictions of gaseous adsorption properties. Chapter 4 also focuses on machine learning study for predicting adsorption properties, with convolutional neural network that self-learns useful features for accurate predictions. Lastly, Chapter 5 summarizes the preceding chapters and provides directions for future work.

Chapter 2. Molecular Model Development for Energy-related Small Molecules

2.1 Systematic Molecular Model Development with Reliable Charge Distributions for Gaseous Adsorption in Nanoporous Materials

E. H. Cho and L. Lin, *J. Mater. Chem. A*, 2018, **6**, 16029 **DOI:** 10.1039/C8TA03737J-Reproduced by permission of The Royal Society of Chemistry

2.1.1 Introduction

Successful screenings of nanoporous materials using molecular simulations rely on the accuracy in describing intermolecular interactions between the gaseous adsorbates and the adsorbents.[29-31] Particularly, an essential component is to have molecular models that are capable of accurately presenting the gas molecules of interest. To date, there have been numerous models reported for small gaseous molecules, and these models have been demonstrated to reproduce some experimental aspects (e.g., liquid density, vapor-liquid equilibrium (VLE), and etc.). However, specific attempts to accurately represent the electrostatic potential (ESP) surface surrounding the gas molecule have not yet been the focus of their development. This is especially important when the adsorption properties of gases are studied in adsorbents where Coulombic interactions are dominant, as in MOFs with coordinately unsaturated open-metal sites that can provide strong affinity to polar molecules or those with strong quadrupole moments.[32] In this book, we demonstrate below that hydrogen sulfide (H_2S) models, for instance, with a relatively poor ESP representation can lead to inaccurate predictions of H_2S binding energy and geometry adsorbed in Mg-MOF-74, one of the most well studied open-metal-site MOFs. Additionally, the parametrization scheme of molecular models reported to date generally does not exhaustively consider all possible model parameters. For instance, using H_2S again as an example, partial charge of the sulfur atom was directly set to zero in many of the available models.[33,34] Such choice in fact appears reasonable considering the computational cost; however, it may also seem arbitrary.

To this end, we introduce in this work a systematic, robust, and computationally-inexpensive approach to develop accurate molecular models by conducting an extensive search of all possible model variables. Specifically, we propose to, for the first time, determine the number of sites needed, along with their optimized locations and charge distributions based on the ESP computed from first-principles density functional theory (DFT) calculations. These variables are determined by a computationally inexpensive scheme to reproduce the DFT-computed ESP, while the remaining variables, primarily the Lennard-Jones (LJ) parameters for van der Waals (vdW) interactions, are subsequently parameterized by fitting to the VLE data that involves significantly more expensive calculations. Not only do models developed with this method can ensure an accurate presentation of the ESP surrounding the molecule, the overall degrees of freedom in the model development can be decoupled and therefore result in orders of magnitude lower computational costs, as opposed to a simultaneous optimization process for all variables. We note that, although periodic DFT is adopted as the main approach in the proposed methodology, a variety of other non-periodic high-level quantum chemical methods can also be used. While this methodology is anticipated to be generalized enough to apply for varying small molecules, we particularly focus on H_2S as a proof of concept. As the removal of H_2S is of utmost importance owing to its harmfulness to our environment and humankind's health, Occupational Safety and Health Administration has currently set a permissible H_2S exposure limit to 10 ppm for general industry.[35] Computational studies for the discovery of optimum nanoporous adsorbents for desulfurization are therefore essential, however such studies remain largely limited by lack of accurate H_2S models.

Since 1986, there have been numerous molecular models proposed for H_2S. The first H_2S model was proposed by Jorgensen,[36] a 3-sites model developed based on the OPLS force field[37] to fit to a dipole moment of 2.1 D. Although H_2S has only three atoms, there have been attempts to better describe the molecule by introducing additional massless sites that only contain point charges, resulting in 4-sites or 5-sites models. In 1989, for example, Forester et al. proposed two H_2S models, one 4-sites and one 5-sites model, where point charges were fitted to reproduce the dimer structure.[38] Afterwards in 1997, Kristóf and Liszi also developed a 4-sites model by fitting point charges and LJ parameters to experimental VLE.[39] In 2003, Nath proposed another 3-sites model with the geometry and point charges of H_2S determined from quantum mechanical calculations.[40] Unlike previous models, LJ parameters were assigned to both the sulfur and hydrogen atoms. In 2013, Gutiérrez-Sevillano et al. further proposed one 3-sites model and two 5-sites models, where the point charges for their models were fitted to the same dipole moment as the model of Kristóf and Liszi or the experimental value (i.e. 1.43D and 0.97D, respectively).[33] The most recent models for H_2S are the TraPPE models by Shah *et al.*, where charges were fitted to reproduce the experimental liquid phase relative permittivity.[34] Shah *et al.* proposed several 3-sites and 4-sites models, where vdW interactions were assigned to only sulfur atoms or both the sulfur and hydrogen atoms.

While these models are able to reproduce some experimental properties, they are not developed with the intention to accurately represent the ESP surface surrounding H_2S. Models with comparatively poor representation of ESP can lead to adsorptive properties largely deviated from the experimental or quantum mechanical results. We present in this work that H_2S models with accurate ESP representation developed using our methodology show huge improvements in the predictions of adsorption phenomena in MOFs and an all-silica zeolite. In this section, we first discuss our proposed methodology. Subsequently, computational details regarding H_2S model development are given. In the results and discussion section, we introduce our developed H_2S models with experimental validations. In this study, as the primary goal is to discover new adsorbent materials for separation, adsorption properties in nanoporous adsorbents is the main focus for the validations. H_2S adsorption properties in all-silica zeolite CHA and in two representative MOFs, open-metal site Mg-MOF-74 and MIL-47 (V), are computed using our models. These results are compared with available experimental data or results computed using DFT. Additionally, a variety of bulk liquid properties other than VLE including radial distribution function (RDF), density, and self-diffusivity are also calculated and compared with available reference data. A comprehensive assessment and comparison of our models with those available in the literature is carried out as well. Finally, with accurate H_2S models at our disposal, H_2S adsorption properties in more than 160 siliceous zeolites included in the International Zeolite Association (IZA) are studied to identify promising candidates for H_2S removal.[11]

2.1.2 Methodology

Overview

Figure 2.1 (a) illustrates the procedure of our methodology, which consists of the following steps:

- ***Geometry and ESP surface:*** employ *ab initio* calculations (i.e., DFT) to determine the geometry of the molecule of interest, while the ESP surrounding the molecule is also computed simultaneously.

- ***Number and location of massless sites and charge distributions:*** consider models with varying numbers of massless sites. For a given number of pseudo-sites, the location of the sites and the corresponding charge distributions are optimized to best reproduce the DFT-computed ESP surface from the previous step. Only models that can reproduce the ESP surface to a certain degree will be considered to proceed further.

- ***Lennard-Jones potential:*** determine Lennard-Jones (LJ) parameters for physical sites by fitting to the experimental vapor-liquid equilibrium (VLE) data.

- ***Model validation:*** validate the developed models by comparing their adsorptive and bulk properties with available experimental or quantum mechanical references.

Details for each step are described below using the development of H_2S models as an example.

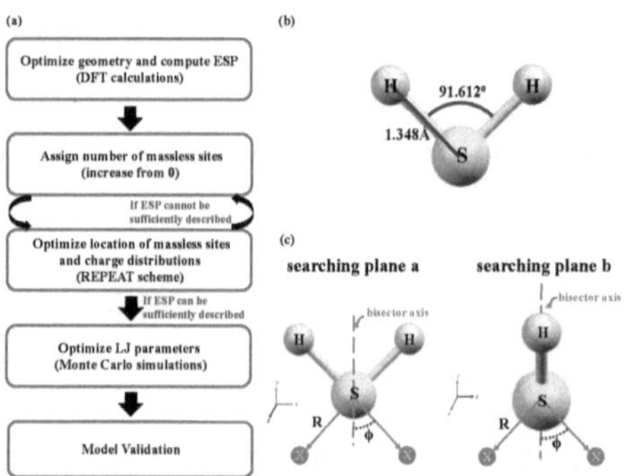

Figure 2.1 (a) Flow diagram of the methodology for molecular model development. (b) DFT-optimized geometry of H_2S. (c) Schematic illustration of searching axis/plane of massless sites (denoted as X) for H_2S models. The searching axis is shown as the bisector axis while the searching planes a and b are the H_2S plane and the plane perpendicular to the H_2S one, respectively.

2.1.2.1 Geometry and ESP surface

To systematically define the geometry of molecules for the model development, periodic density functional theory (DFT) calculations are employed to optimize the geometry of an isolated molecule. DFT has been proven to capture many details of electronic structures as shown in many of the previous studies, thus offering an accurate means to determine the structure of gas-phase molecules.[27,41] The ESP surface surrounding the DFT-optimized molecule is also obtained simultaneously from the calculations and subsequently used in the next step. For H_2S, the DFT-optimized structure can be seen in Figure 2.1 (b). As noted previously, although DFT is adopted as the primary tool, other high-level cluster-based quantum chemical methods can also be used in this step and the following steps.

2.1.2.2 Number and location of massless sites and charge distributions

Number of massless sites: Massless sites that only possess Coulombic interactions have been commonly included in a variety of models as in some of the available H_2S models introduced previously. In our approach, we consider the number of massless sites as a parameter, by varying the numbers of massless sites (e.g., 0, 1, 2, and 3), starting from zero and then sequentially include more massless sites until a reasonable ESP representation is achieved (see Figure 2.1(a) for the flow diagram). We note that, although having more massless sites may better describe the ESP, this also results in a model requiring a substantially higher computational cost while being employed in molecular simulations. It is therefore preferred to have as few as possible massless points as long as the model can reasonably resemble the ESP surface.

Location of the massless sites and charge distributions: For a given number of massless sites, their location and the overall charge distribution of the model are optimized to best resemble the ESP surface. To preserve the symmetry of the ESP surface for some molecules, certain constraints have to be imposed while exploring the position of massless sites. For H_2S, one can only consider massless sites sitting on specific axis (denoted as searching axis) and/or in certain planes (denoted as searching plane), as schematically illustrated in Figure 2.1(c). The searching axis and/or planes for each number of massless sites of the H_2S models are summarized as followings. For clarity, the models are named hereafter as "(number of sites)S-(searching plane for the massless sites)", e.g., 5S-a refers to a 5-sites model with 2 massless sites in the searching plane a.

(1) 3-sites model: No massless site.

(2) 4-sites model: One massless site along the bisector axis

(3) 5-sites model:

      a.   Two massless sites in the plane a. (5S-a)

      b.   Two massless sites in the plane b. (5S-b)

(4) 6-sites model:

      a.   Two massless sites in the plane a and one site along the bisector axis. (6S-a)

      b.   Two massless sites in the plane b and one site along the bisector axis. (6S-b)

For each of the models except the 3-sites model, an extensive grid-search is carried out to examine all possible locations of massless sites and compute the corresponding charge distribution of the model. Specifically, to determine the charge distribution of each possibility, we use the REPEAT scheme for optimizing atomic charges to best reproduce the ESP surface.[42] The REPEAT scheme is a numerical approach to designate the best set of atomic charges with a minimized error of ESP computed by point charges using the

Ewald summation technique relative to that computed by DFT. The error is quantified as the so-called relative root-mean square (RRMS), which is defined as

$$\text{RRMS} = \sqrt{\frac{\sum_{grid}(\phi_{QM}(\vec{r}_{grid}) - \phi_q(\vec{r}_{grid}))^2}{\sum_{grid}\phi_{QM}(\vec{r}_{grid})^2}} \qquad (2.1)$$

where ϕ_{QM} and ϕ_q denote the electrostatic potential computed by DFT and point-charge methods, respectively. A smaller RRMS value indicates a better ESP representation. Finally, the optimized location of massless sites and the corresponding charge distribution of the model can be chosen as the one with a global minimum in RRMS. As an example, Figure 2.2 and Figure 2.3 illustrates the RRMS distribution as well as the charge distribution as a function of the location of massless sites for the 5S-a model and 5S-b model, respectively. For the grid-search carried out here, a sufficiently small step size of $5°$ and 0.07 Å are chosen for angle, ϕ, and location, R, as shown in Figure 2.1 (c), respectively. Such choice has led to RRMS value varying negligibly between neighboring grid points near the global minimum (i.e., ΔRRMS < 0.01).

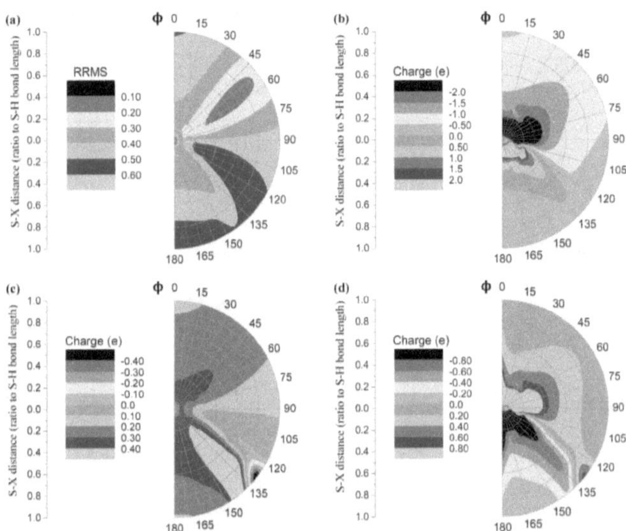

Figure 2.2 H$_2$S 5S-a model (a) RRMS distribution and the corresponding charge for (b) sulfur, (c) hydrogen, and (d) massless site as a function of the massless sites location, R and ϕ (see Figure 2.1(c)). The S-X distance (denoted as R in Figure 2.1(c)) is plotted as the ratio with respect to the S-H bond length.

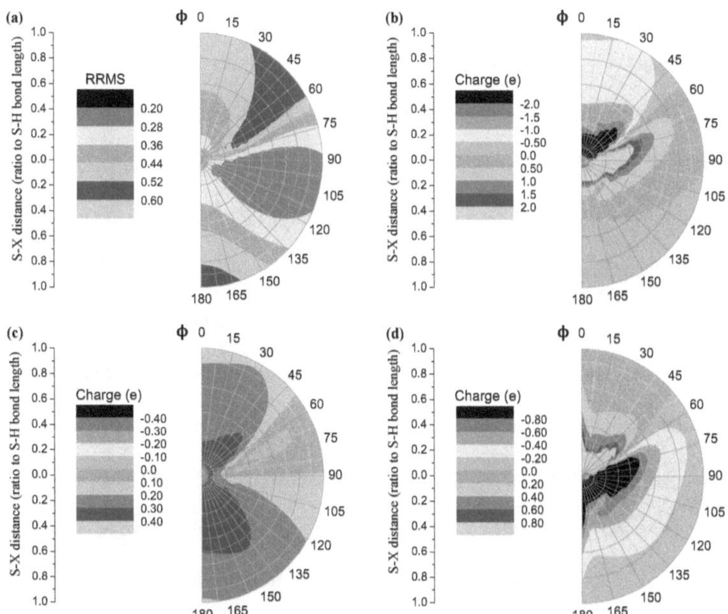

Figure 2.3 H$_2$S 5S-b model (a) RRMS distribution and the corresponding charge for (b) sulfur, (c) hydrogen, and (d) massless site as a function of massless sites location, R and ϕ (see Figure 2.1 (c)). The S-X distance (denoted as R in Figure 2.1(c)) is plotted in the figure as the ratio with respect to the S-H bond length.

2.1.2.3 Lennard-Jones Parameter Determination

For models that have a sufficiently small RRMS, the Lennard-Jones (LJ) parameters of physical sites (e.g., S and H of H$_2$S) are then fitted to reproduce experimental VLE data.

In this optimization step, parameters search is conducted in a straightforward manner, where a grid-search is carried out for epsilon and sigma values. For each set of possible parameters, the corresponding VLE is computed using Gibbs ensemble Monte Carlo (GEMC) simulation techniques. The optimized parameters are the ones that result in the least error relative to the experimental VLE data. To ensure that the mathematically-driven optimized force field is physically sensible, the corresponding critical temperature is also computed and compared with the experimental data. A criterion is set for the difference in the critical temperature to be within 3 %. For those models that meet all the criteria, final model validation is further proceeded. For H_2S model developments, we firstly develop models with only LJ parameters assigned for sulfur (i.e., treat hydrogens as non-vdW sites). Treating the biggest atom of a molecule (i.e., sulfur) as the only vdW site is a commonly adopted assumption as in most of the previously proposed models. At this point, it is important to highlight again that, in our approach, the Coulombic and dispersion parameters are decoupled. This can not only ensure the ESP of the molecule to be described accurately but also avoid including all possible parameters in the fitting to VLE data that requires notably more expensive GEMC calculations. As a result, orders of magnitude lower computational cost are expected, and an extensive search over all possible model parameters in a systematic manner can be achieved.

2.1.2.4 Model Validation

With the molecular models developed from the previous steps, a variety of properties including those in adsorbed and bulk liquid phases can be investigated for validations. Both

Monte Carlo (MC) and molecular dynamics (MD) simulations are carried out for such purpose.

2.1.3 Computational Details

Computational approaches for predicting a variety of properties using DFT, Monte Carlo, and molecular dynamics simulations are given below.

2.1.3.1 Density functional theory (DFT) calculations

DFT calculations are used to optimize the structure of H_2S molecules in gas phase and obtain the correspondent ESP surface. Specifically, a H_2S molecule is placed in a cubic cell with a cell dimension of 15 Å. In these calculations, all atomic positions are relaxed until the residual forces on every atom are smaller than 0.02 eV/Å using the Quantum ESPRESSO[43] implementation with the Perdew-Burke-Ernzerhof (PBE) exchange-correlation functional and Troullier-Martins type norm-conserving pseudopotentials. The kinetic cut off is set to 80 Rydberg.

2.1.3.2 Molecular Simulations

Vapor liquid equilibrium (VLE) and critical properties: MC simulations in the Gibbs ensemble (GEMC) implemented in the RASPA package[44] are carried out to compute the VLE properties of a given H_2S model for a temperature range from 220 K to 340 K with an increment of 10 K. A grid-search optimization is carried out with step sizes of 5 K and

0.01 Å for the epsilon and sigma values, respectively. In these calculations, the probabilities for the MC moves are 5.88 % for volume swap, 11.76 % for translation, 11.76 % for rotation, 23.53 % for reinsertion, and 47.06 % for particle swap.

The critical temperature and density are obtained by fitting the VLE results to the scaling law (Equation 2.2)[45] and the law of rectilinear diameters (Equation 2.3)[46], respectively.

$$\rho_l - \rho_g = B(T_c - T)^\beta \tag{2.2}$$

$$\frac{\rho_l + \rho_g}{2} = \rho_c + A(T - T_c) \tag{2.3}$$

where ρ is density with subscripts l and g denoting the liquid and gas phase, respectively and β is universal critical exponent. VLE data from GEMC simulations with temperature greater than 298 K (0.8 T_c) are employed for the fitting. β in the scaling law is also included as a parameter instead of being assigned as 0.325 for three-dimensional fluids.[45] In consistent with other works, difference in β is found to have negligible effects on the determined critical properties.[33]

Adsorption isotherms: MC simulations in a grand canonical ensemble (GCMC simulations) are employed to compute the adsorption isotherms of gases in nanoporous materials. Millions of MC moves, including translational, rotational, reinsertion, and swap (insertion/deletion) are carried out to obtain statistically accurate ensemble average. The relative probabilities for the aforementioned moves are set to 12.5 %, 12.5 %, 25 %, and 50 %, respectively.

Binding geometry and energies: Binding geometry and energy of H_2S in structures of interest are computed using MC simulations in a canonical ensemble (NVT simulations). A single H_2S molecule is placed in the simulation box. Multiple replica of NVT simulations are carried out at a low temperature of 0.1 K and each of them with millions of MC attempts of translation, rotation, or reinsertion (probability of 25 %, 25 %, and 50 %, respectively). In such a low temperature condition, the system would evolve to a configuration with a lower energy. The configuration with the lowest energy amongst all configurations is identified as the binding geometry.

Henry Coefficients and Heats of Adsorption: Henry coefficients and heats of adsorption of H_2S in the infinite dilute condition are calculated using the Widom particle insertion method.[30] A guest molecule is inserted randomly into the structure where the corresponding Rosenbluth factor and energy are computed.

Liquid density, radial distribution function, and diffusivities: MD techniques implemented in the open-source LAMMPS package[47] are employed to compute the bulk properties of H_2S. The initial configuration of the simulation system consists of 448 H_2S molecules in a cubic box with each dimension of 40 Å. This initial system is firstly equilibrated in a canonical ensemble for 0.1 ns using the Nosé-Hoover thermostat with a damping factor of 100 timesteps (i.e., 100 fs). Subsequent equilibration is done in an isothermal-isobaric ensemble, regulated by the Nosé-Hoover barostat and thermostat with a damping factor of 1,000 and 100 timesteps, respectively, for another 0.1 ns. Finally, a production run in the isothermal-isobaric ensemble with a duration of 2 ns is conducted in which the density and RDF are computed. MD simulations are also used to compute the self-diffusivity of liquid H_2S. The same equilibration runs are done as described above, followed by a production run in a microcanonical ensemble with a duration of 2 ns in which the mean square displacement of the molecules as a function of time is obtained.

Structures and force field parameters: The crystallographic structure data of all zeolites studied in this work are taken from the International Zeolite Association.[11] The force field parameters for the zeolite framework are adopted from CLAYFF.[48] For Mg-MOF-74, the DFT-optimized structure and DFT-derived charges from a previous study are used.[41] The LJ parameters of the framework atoms are assigned from the Universal force field.[49] For MIL-47(V), experimental structure from Barthelet *et al.* is used.[50] Point charges for the framework atoms are taken from a previous simulation work of Gutiérrez-Sevillano *et al.*[33] For the LJ parameters, two different sets are used; all the atoms are treated with DREIDING force field except the metal from UFF or except the metal and oxygen from UFF,[49,51] as proposed respectively by Gutiérrez-Sevillano *et al.*[33] and Hamon *et al.*[52] The Lorentz-Berthelot mixing rule is applied to estimate the LJ parameters between dis-similar atoms. In all of the simulations carried out in this study, LJ potential is truncated and shifted to zero at a cutoff radius of 12 Å, while Coulombic interactions are computed using a long-range solver with a relative error of 10^{-6}. Specifically, the Ewald summation technique and the particle-particle particle-mesh method are adopted respectively in MC and MD simulations. We note that, for pores or channels inaccessible to H_2S (kinetic diameter of 3.64 Å) in zeolites, blocking pockets identified from a previous study using N_2 (kinetic diameter of 3.6 Å) as a probe molecule are used.[53]

2.1.4 Results and Discussion

2.1.4.1 Model Development

We apply the methodology for the H_2S model development. The DFT-optimized geometry of H_2S as shown in Figure 2.1 (b) has a H-S length of 1.348 Å and a H-S-H angle of 91.612°, which are comparable with that used in all models reported in the literature. To determine the number of massless sites being included, as introduced previously, the value of RRMS is used as the objective criteria. It is found that models with 0 and 1 massless sites (i.e., 3-sites and 4-sites models) may unable to adequately represent the ESP surface, as indicated by the relatively large RRMS values of 0.6 and 0.35, respectively. Including two massless sites (5-sites models) instead can substantially reduce the RRMS to approximately 0.2. Having three massless sites (6-sites models) can further lower the RRMS but the improvement is marginal. The optimized model parameters for these H_2S models with 2 or 3 massless sites are summarized in Table 2.1. The corresponding RRMS values and dipole moments are summarized in Table 2.2. Interestingly, all of these 5 and 6-sites models have nearly the same dipole moment that is close to the experimental value of 0.97D, suggesting that the point charges assigned from the REPEAT scheme have electronic distribution analogous to the real molecules. For comparison, Table 2.3 summarizes the dipole moment of previously reported H_2S models. Except for the 5Sd model by Gutiérrez-Sevillano *et al.*, where their point charges are fitted to the experimental dipole moment, none of the previous models resemble the experimentally-determined dipole moment. This is also the case for the 3-sites model by Nath that have point charges assigned by *ab initio* methods. To ensure accurate description of the ESP surface, 5- and 6-sites models that possess lower

RRMS values are desired and considered in this study for further parameterization and studies. Specifically, 5-sites models are of particular focus in this work for its computational efficiency, as compared to 6-sites models. We note that, although H_2O molecules have a similar molecular geometry as H_2S, there have had a variety of successful 3-sites and 4-sites H_2O models.[54,55] Unlike H_2O, however, our results here indicate that H_2S requires more massless points, which may be attributed to their bigger molecular size and less electronegative sulfur atom.

Table 2.1 Geometry and parameters of the (a) 5-sites and (b) 6-sites models developed in this work. X denotes the massless sites, where for the 6-sites models, X_1 is the massless site sat along the bisector axis and X_2 represents those two massless sites located in the searching plane a or b. The definition of R and ϕ can be found in Figure 2.1 (c).

(a)

	Geometry		Charge			LJ parameter	
	R (Å)	ϕ (°)	qS (e⁻)	qH (e⁻)	qX (e⁻)	ε_S (K)	σ_S (Å)
5S-a						270	3.76
5S-a-D	0.8764	55	-1.152	0.268	0.308	255	3.48
5S-a-U						180	3.58
5S-b						270	3.75
5S-b-D	0.5393	100	1.106	0.176	-0.729	260	3.48
5S-b-U						182.5	3.57

(b)

	Geometry			Charge (e⁻)				LJ Parameter	
	R_{S-X_1} (Å)	R_{S-X_2} (Å)	ϕ (°)	qS	qH	qX_1	qX_2	ε_S (K)	σ_S (Å)
6S-a	0.6067	0.6741	60	-1.858	0.228	0.252	0.575	270	3.76
6S-b	0.4719	0.5393	80	1.981	0.177	-0.636	-0.8495	270	3.76

Table 2.2 Values of RRMS and dipole moment with the predicted critical properties for the developed models in this work. The experimental reference values are from NIST.[56,57]

	RRMS	Dipole Moment (D)	T_c (K)	ρ_c (kg/m³)
Exp.	-	0.97[56]	373.1[57]	347.3[57]
5S-a	0.163	0.93	367.2	357.5
5S-b	0.218	0.93	364.0	360.8
6S-a	0.154	0.93	365.7	361.9
6s-b	0.167	0.92	363.9	347.9

Table 2.3 Dipole moment computed by the available (a) 3-sites, (b) 4-sites, and (c) 5-sites H_2S models reported in the literature.[33,34,36,38–40]

	Dipole Moment (D)
(a) 3-sites	
Jorgensen[36]	2.10
Nath[40]	1.13
Gutiérrez-Sevillano et al.[33]	1.43
Shah et al. (TraPPE 3-1)[34]	1.12
Shah et al. (TraPPE 3-3)[34]	1.25
(b) 4-sites	
Forester et al.[38]	1.34
Kristóf and Liszi[39]	1.43
Shah et al. (TraPPE 4-1)[34]	1.32
Shah et al. (TraPPE 4-3)[34]	1.27
(c) 5-sites	
Forester et al.[38]	1.41
Gutiérrez-Sevillano et al. (5S)[33]	1.44
Gutiérrez-Sevillano et al. (5Sd)[33]	0.97

Figure 2.4 shows the VLE of the 5S-a, 5S-b, 6S-a, and 6S-b models with their LJ parameters of sulfur fitted to the reference experimental VLE data from NIST.[57] As also summarized in Table 2.2, the corresponding critical temperature and density obtained are also found to agree well with that determined experimentally. Interestingly, the LJ parameters of models developed in this work are very similar (i.e., epsilon and sigma of sulfur to be ~270K and ~3.76Å, respectively), although their geometry and charges are drastically different. This can be attributed to that each of our models has equally good representations of the ESP surface. The H_2S models developed in such a way to reproduce the ESP surface offer accurate Coulombic interactions between the molecules and therefore, LJ parameters values are alike to each other. In other words, models with their electronic distributions different from each other would possibly result in a set of dissimilar LJ parameters. Indeed, previously published models that have LJ parameters also fitted to VLE show very different values of those parameters. For example, the 5-sites model with dipole moment of 0.97D (denoted as 5Sd) developed by Gutiérrez-Sevillano *et al.*, has a notably higher epsilon value of 310 K.

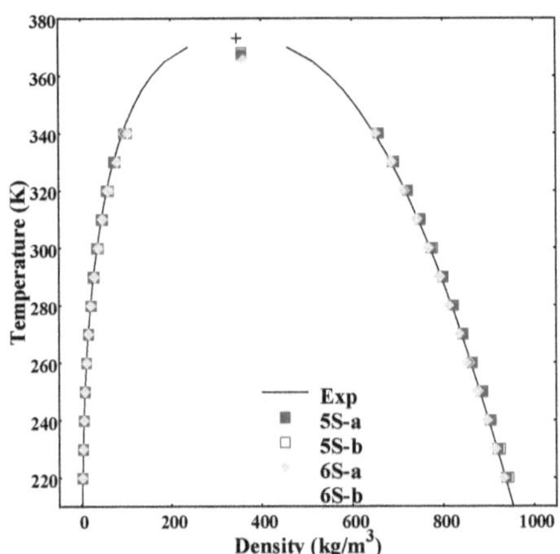

Figure 2.4 Vapor liquid equilibrium computed by the models developed in this work. The experimental reference from NIST is included for comparison.[57]

2.1.4.2 Model Validation – Adsorption Properties

Metal-organic Framework: Mg-MOF-74

Toward the goal of employing these models in the discovery of promising adsorbents, we firstly investigate their predictions in the binding energy and geometry of H_2S in the open-metal-site Mg-MOF-74 material as a good representation of the molecular ESP is expected to be critically important in these MOFs. To date, H_2S physisorption experimental measurements in Mg-MOF-74 remain unavailable. However, the binding energy and geometry of H_2S in Mg-MOF-74 have been studied using DFT by Lee *et al.*, [27] which can serve as accurate references to validate our developed molecular models. As expected and can be seen below from our results, binding energy and adsorption uptake (i.e., in particular at a low-pressure region) are highly correlated. Therefore, this implies that model that results in a binding energy comparable to the DFT-computed one may possibly also be accurate in their predictions of adsorption.

Binding energy and binding geometry predicted with models developed in this work and some selected available models are summarized in Table 2.4. A comprehensive summary of the results using available models reported to date can be seen in Table 2.5. Tables 2.4 and 2.5 show our models achieve significant improvements compared to previously reported models as the predicted binding energies with our models resemble DFT results accurately. For instance, our models predict a binding energy ranging from 5262 to 5703 K, comparable to the DFT-predicted value of 5773 K. We note that the computed binding energies by these classical models may exist uncertainties due to the choice of framework LJ parameters and the choice of approximation for pair-wise cross-atom interactions (i.e., mixing rule). Additionally, for the DFT-computed energy, uncertainties may exist due to the choice of vdW functional. For this reason, all of 5S-a, 5S-b, 6S-a, and 6S-b models are deemed accurate enough to be employed in adsorption simulations specifically in open-metal sites MOFs. By contrast, most of the models in the literature predict a notably lower binding energy (e.g., TraPPE 3-1 model by Shah *et al.* predicts 3496 K, see Table 2.5). With respect to the corresponding binding geometry, the binding distance (i.e., between S (H_2S) – Mg (framework)) and the binding angle (i.e., between the plane of H-S-H and the plane of one Mg and four O of the framework, referred as Mg-O$_4$) are computed. A schematic representation can be seen in Figure 2.5. Table 2.4 shows that the binding distance associated with our 5S-a, 5S-b, 6S-a, and 6S-b models are also all in excellent agreement with the DFT-computed distances (i.e., ~3.00 Å), and the predicted binding orientations are reasonably similar to the parallel orientation determined from DFT (i.e., close to 0°) as well. Moreover, these results regarding the binding geometry predictions by our models appear to be significantly improved compared to other models in the literature. For instance, TraPPE 3-1 model predicts a much larger binding distance of 3.23 Å and a perpendicular binding geometry with a binding angle of 86°. For a clear illustration, the figures of binding geometry computed by DFT, 5S-a, and TraPPE 3-1 are shown in Figure 2.6 (a), (b), and (c), respectively. A more detailed comparison over our models and all possible models in the literature can be found later.

Although notable improvements have been achieved, all of our models predict a negative binding angle. This indicates that the hydrogen atoms are a little tilted toward the Mg-O$_4$ plane, whereas a binding angle of almost 0° is predicted by DFT. We hypothesize that this may be attributed to the non-physical hydrogen atoms of the H_2S models (i.e., no LJ parameters). Since the open-metal sites MOFs provide strong inductive interactions with the guest molecules, there can be strong attractive Coulombic interactions between the positively charged hydrogen atom of H_2S and the negatively charged oxygen atom of the framework. However, as the hydrogen atom approaches the oxygen of the framework, there should be repulsion between them due to the electron cloud overlapping but this repulsive interaction is not included in these H_2S models.

Table 2.4 Binding energy and geometry of H_2S adsorbed in Mg-MOF-74 computed by the proposed models in our work. DFT results from Lee *et al.* are also included for reference.[27] The percentage of the Coulombic energy to the total binding energy is also given in this table.

	Binding Energy (K)	Coulombic Energy (%)	Binding Distance (Å)	Binding Angle (°)
DFT[27]	5773	-	2.97	+1.68
5S-a	5262	69	2.98	-34.85
5S-b	5703	75	2.93	-20.32
6S-a	5270	69	3.00	-26.74
6S-b	5522	70	2.97	-8.28
5S-a-D	5872	65	2.84	-7.70
5S-a-U	5520	62	2.87	+14.27
5S-b-D	6603	75	2.79	-8.57
5S-b-U	6185	71	2.81	+7.35

Table 2.5 Binding energy of H_2S adsorbed in Mg-MOF-74 computed using the available (a) 3-sites, (b) 4-sites, and (c) 5-sites models reported in the literature.[33,34,36,38–40] The percentage of the Coulombic energy to the total binding energy is also given in this table.

	Binding Energy (K)	Coulombic Energy (%)	Binding Distance (Å)	Binding Angle (°)
(a) 3-sites model				
Jorgensen[36]	3852	82	3.03	+68.54
Nath[40]	3416	29	3.24	+87.92
Gutiérrez-Sevillano et al.[33]	3771	35	3.17	+80.56
Shah et al. (TraPPE 3-1)[34]	3496	28	3.23	+86.45
Shah et al. (TraPPE 3-3)[34]	3873	30	3.08	+29.85
(b) 4-sites model				
Forester et al.[38]	4528	66	3.03	-47.73
Kristóf and Liszi [39]	3622	34	3.23	+26.04
Shah et al. (TraPPE 4-1)[34]	4508	68	3.12	-50.68
Shah et al. (TraPPE 4-3)[34]	3917	30	3.05	+15.99
(c) 5-sites models				
Forester et al.[38]	5889	92	2.81	-7.57
Gutiérrez-Sevillano et al. (5S)[33]	3849	34	3.20	+47.83
Gutiérrez-Sevillano et al. (5Sd)[33]	3507	24	3.24	+51.44

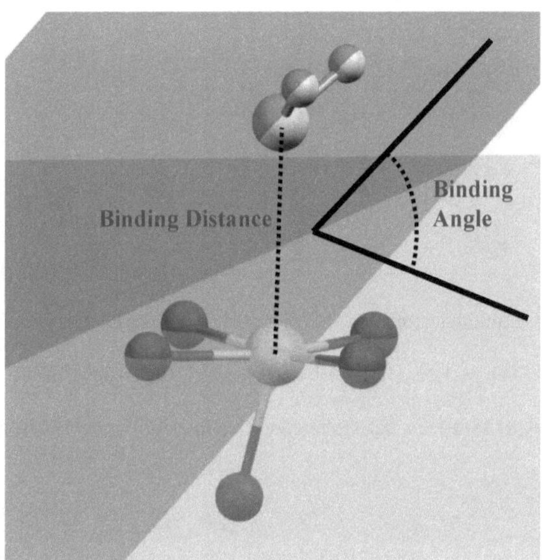

Figure 2.5 Schematic representation of the binding distance (between S (H_2S)-Mg (framework)) and binding angle (between H-S-H (H_2S) plane and the MgO_4 (framework) plane) of H_2S adsorbed to the open-metal sites of Mg-MOF-74. Color codes: yellow-sulfur, grey-hydrogen, green-magnesium, and red-oxygen.

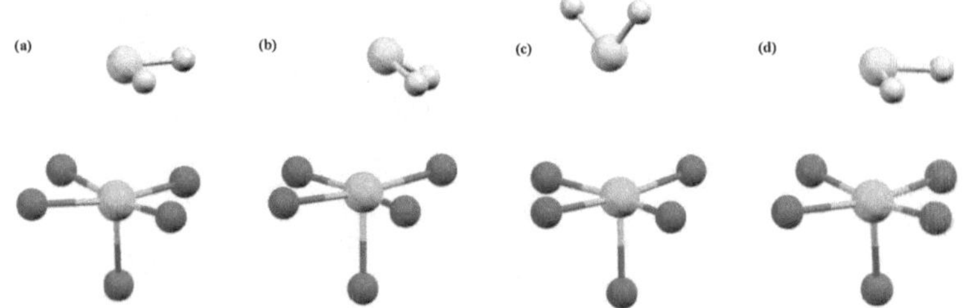

Figure 2.6 Binding geometry of H_2S adsorbed in open-metal sites Mg-MOF-74, computed with (a) DFT by Lee *et al.*, [27] (b) 5S-a model, (c) 3-sites TraPPE 3-1 model by Shah *et al.*,[34] and (d) 5S-a-D model. Color codes: yellow-sulfur, grey-hydrogen, green-magnesium, and red-oxygen.

To verify this hypothesis, we also attempt to include LJ parameters for the hydrogen atoms in our 5S models. As a proof of concept and for simplicity, instead of directly fitting 4 parameters (i.e., ε_S, σ_S, ε_H, and σ_H) simultaneously to the experimental VLE, hydrogen parameters from either Universal force field (UFF)[49] or DREIDING force field are adopted.[51] The same optimization procedure is then followed to parameterize the LJ parameters of the sulfur atom. These 5S H_2S models are denoted with an additional specification for the inclusion of hydrogen LJ parameters (e.g., 5S-a-U (UFF) or 5S-a-D (DREIDING force field)). Their parameters are summarized in Table 2.1 (a) and the corresponding VLE curves are shown Figure 2.7.

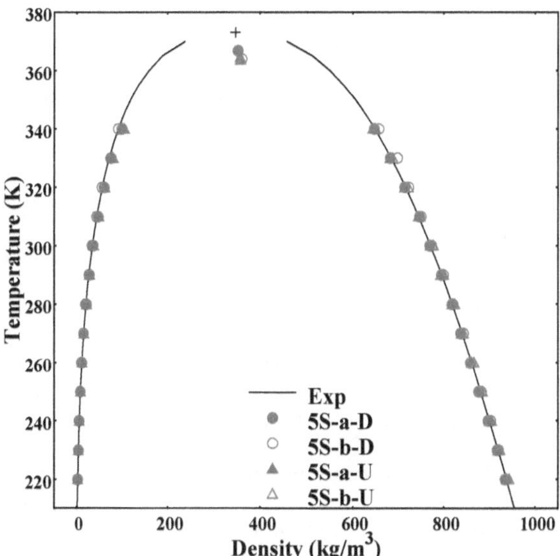

Figure 2.7 Vapor liquid equilibrium computed using the developed models that include hydrogen vdW parameters. The experimental reference from NIST is included for comparison.[57]

The binding geometry computed using these models with hydrogen LJ parameters clearly show that the hydrogen atoms now experience repulsive LJ interaction, as indicated by the more positive values of their respective binding angle (see Table 2.4). Figure 2.6 (d) visually show that the DFT-predicted binding angle can be well captured by 5S-a-D. However, the corresponding binding distances are now shorter than that from DFT. This may be due to that the sulfur atom of H_2S becomes less repulsive from the framework as the LJ interaction energy is now distributed throughout the whole molecule, as can be inferred from the smaller values of σ and ε assigned to the sulfur atom. However, it should be noted that the LJ parameters of hydrogens are not optimized, but rather assigned directly from generic force fields. A more detailed development to include the hydrogen LJ parameters into the parameterization may further improve the accuracy of the models, but such efforts might be somewhat unnecessary since 5S-a and 5S-b models can already decently reproduce the binding energy and geometry.

Figure 2.8 shows the adsorption isotherms of H_2S in Mg-MOF-74 predicted with our 5-sites models and available models in literature. In Figure 2.8 (a), the differences in loading between our developed models are minor (i.e., two-fold difference in loading at 5 kPa while no observable difference at a higher pressure) given that a small difference in force field can be highly amplified in the uptake predictions for structures with coordinately unsaturated metals that can interact strongly with polar guest molecules, especially at a low-pressure region. As is well known, the adsorption loading at low-pressure region and the binding energy are highly correlated. From our results, 5S-b-D and 5S-b-U show the highest adsorption loading followed by 5S-a-D and 5S-a-U, and then by 5S-b, and lastly by 5S-a, which agree with the general trend in the predicted binding energy. Unlike our models, loading predictions made by other available models in the literature can vary tremendously, by as much as 15 times at 5 kPa. (5-sites model by Forester *et al.* vs. 3-sites model by Jorgensen). Furthermore, even in a relatively higher-pressure range, observable differences can be identified for the available models. For instance, at a pressure of 1 bar, prediction from 5-sites model by Gutiérrez-Sevillano *et al.* (denoted as 5S) shows a loading

of 7 mol/kg larger than that predicted from the 3-sites model by Jorgensen.

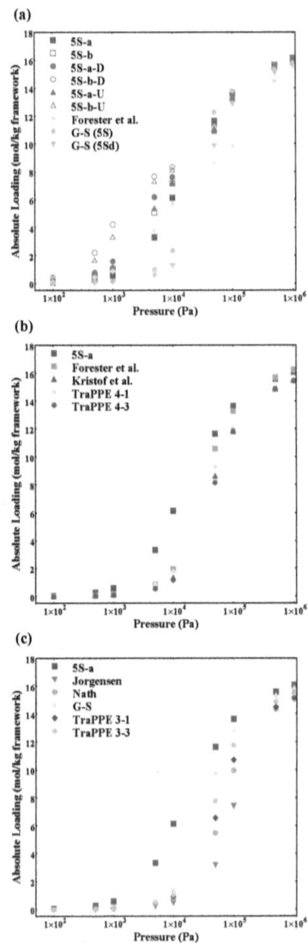

Figure 2.8 Adsorption isotherms of H_2S in Mg-MOF-74 using available (a) 5-sites, (b) 4-sites, and (c) 3-sites models, including those developed in this work and reported in the literature.[33,34,36,38–40] Predictions by 5S-a model are also included in both (b) and (c) for comparison. For brevity, G-S refers to Gutiérrez-Sevillano *et al.*

From the results discussed above, we find that the binding energy/geometry and adsorption loading largely depend on the number of sites of each model. Generally, models with higher number of sites predict a stronger binding energy, which are much closer to that computed from DFT. This may be attributed to the better representation of ESP surface when more massless sites are used, consistent with our findings that at least two massless sites are needed to develop a physical H_2S model. Similarly, binding geometry also show an interesting trend with respect to the number of sites included in each model. In contrast to the DFT predicted results, as noted above, calculations using several 3-sites models in the literature show a perpendicular orientation (see Figure 2.6 (c)) and an overestimation of the S – Mg distance by as large as more than 0.2 Å. Incorporating one massless site to the H_2S molecular model (i.e., 4-sites models) does not show significant improvement in terms of binding geometry in Mg-MOF-74. For instance, binding geometry predictions by TraPPE 4-1 model and 4-sites model by Forester *et al.* have a binding angle of approximately negative 50°. For 3-sites and 4-sites models with hydrogen LJ parameters included (i.e., TraPPE 3-3 or 4-3 models), the predictions for binding geometry show improvement to a certain degree but the binding energy remains underestimated. On the contrary, most of the 5-sites or 6-sites models (i.e., our models and a 5-sites model by *Forester et al.*), clearly show a better prediction in binding geometry as can be seen from Table 2.4 and Table 2.5.

To highlight the important role of the electrostatic potential, the binding energy is decomposed to vdW and Coulombic contributions for all of our models and previously developed models. In a previous work for investigating carbon dioxide adsorbed in Mg-MOF-74, the interaction energy computed by quantum chemical method was decomposed to electrostatic, repulsive, and attractive components by the NEMO decomposition technique.[32] It was found that the interaction between the gas molecule and the framework atoms are largely driven by electrostatic effect. Since H_2S has permanent dipole, it is anticipated that the electrostatic interaction would also dominate the binding energy of H_2S in Mg-MOF-74. Table 2.4 and Table 2.5 show most of those models with 5 or more sites (i.e., our models and a 5-sites model by *Forester et al.*) not only have comparable binding energy to the DFT-computed value but also the dominance of Coulombic interactions. This is not the case though for 5-sites models proposed by Gutiérrez-Sevillano *et al.* For these 5-sites models, the Coulombic energy contributions to the total binding energy are ~30%.

Metal-organic Framework: MIL-47(V)

Adsorption properties of H_2S in another well-studied metal-organic framework, MIL-47(V), are also investigated using H_2S models developed herein and are validated with experimental references.[58] Among experimental H_2S adsorption in metal-organic framework available in the literature, to the best of our knowledge, MIL-47(V) is the only structure which has well-established chemical and topological information.[50,58] For this MOF, there have been some previous computational works conducted to study the adsorption of H_2S.[33,52] For the LJ parameters of the framework atoms, two different sets were proposed in previous works, which subsequently can lead to differences in adsorbate-adsorbent vdW interactions. As will be seen below, vdW interactions between the guest molecule and the framework play a dominant role in MIL-47, unlike open-metal-site Mg-MOF-74. For this reason, simulation results may be more dependent on the choice of adsorbate-adsorbent LJ parameters, especially in relatively confined environment.[59] The first set has parameters assigned from DREIDING for all atom types except for vanadium which is from UFF. The second set only differs from the first one in the oxygen atom, whose parameters also are taken from UFF.[49,51] These sets are denoted as FFO_D and FFO_U, respectively, which the subscripts indicate the source of LJ parameters for the oxygen atom.

Figure 2.9 summarizes adsorption of H_2S in MIL-47(V) predicted with 5S-a and 5S-b models. In the figure, experimental results by Hamon *et al.* are also included for comparision.[52] The results show that our predicted isotherms resemble the experimental measurements accurately. Specifically, as reported by Hamon *et al.*, the experimental isotherm can be classified as a monotonic quasi-type I shape with Henry coefficient of 4.7×10^{-5} mol/kg/Pa. From our calculations, our 5-sites models predict $6.3\sim6.5\times10^{-5}$ mol/kg/Pa and $4.8\sim4.9\times10^{-5}$ mol/kg/Pa using FFO_D and FFO_U, respectively. Although the predicted Henry coefficients can differ from the experimental value by approximately 20 % when FFO_D is used, predictions made by our models can still be deemed in excellent agreement with experiments considering the role of vdW energy and the uncertainties in the LJ parameters for frameworks. This agreement has also suggested that a good representation of H_2S molecules has been achieved. The fact that vdW interactions are significant in MIL-47(V) can also be verified by binding energy calculations with 5S-a and 5S-b models and also some other selected models available in the literature, as summarized in Table 2.6. Table 2.6 clearly demonstrates that the interaction between adsorbate and adsorbent is largely dominated by vdW interaction as opposed to Mg-MOF-74 (i.e., ~80 % vs. ~30 %). Calculations using other available models, such as that by Forester *et al.* or Shah *et al.*, also show the significance of vdW interactions, although their predicted binding energies are slightly different. At this point, it is important to note that the optimization of the adsorbate-adsorbent vdW interactions is beyond the scope of this work and will be an important subject of future studies, especially for studying structures such as MIL-47(V) where vdW interaction between adsorbate and adsorbent is a significant factor. Recently, *ab initio* calculations have been employed for the parameterization of adsorbate-adsorbent interactions as reported in the literature.[32,41,60,61] Accurate molecular models for describing gas molecules are critically needed for this purpose and therefore, our method as presented herein can also largely facilitate these developments.

This comprehensive analysis on the adsorption properties in Mg-MOF-74 and MIL-47(V) highlights the potential of our systematic approach to develop molecular models for accurate description in adsorption properties. We demonstrated that our models can accurately and consistently predict the adsorption properties in Mg-MOF-74, and the results are much improved from other previously developed models. These models are also shown to reproduce the experimental adsorption isotherm of H_2S in MIL-47(V). As will be discussed below, the accuracy of the H_2S models developed by our methodology are further validated for the adsorption in zeolite CHA.

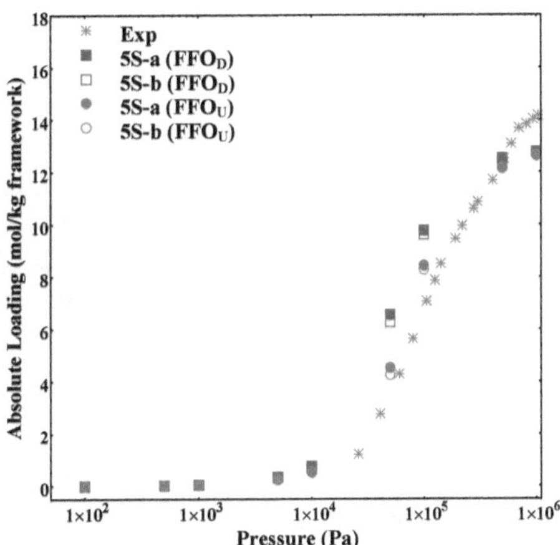

Figure 2.9 Adsorption isotherm of H_2S at 303K in MIL-47(V) predicted by the models developed in this work. Experimental result from Hamon *et al.* is also included.[52]

Table 2.6 Binding energy of H_2S adsorbed in MIL-47(V) computed by the models developed in our work and some selected available models in the literature.[34,38] The percentage of the Coulombic energy to the total binding energy is also given in this table.

	Binding Energy (K)	Coulombic Energy (%)
5S-a (FFO$_D$)	3128	17
5S-b (FFO$_D$)	3117	16
5S-a (FFO$_U$)	2812	21
5S-b (FFO$_U$)	2796	20
Forester *et al*. (5-sites, FFO$_D$)[38]	2627	28
Shah *et al*. (TraPPE 3-1, FFO$_U$)[34]	2828	14

All-silica Zeolite: CHA

The adsorption isotherm of H_2S in all-silica zeolite CHA is also computed using the models developed in this work and previously reported models. Figure 2.10 shows a summary of adsorption isotherms computed using all of models including those reported in the literature. The available experimental data (Maghsoudi *et al*.[62]) is included for comparison, and the results show that the adsorption properties predicted using each of our model is in excellent agreement with the experimental data. To better quantify the comparison, each of these adsorption isotherms is fitted using the Langmuir equation, and the parameters (Henry coefficient and saturation loading, denoted as K_H and Q_{sat}, respectively) can be seen in Table 2.7 for our 5-sites models and the models reported in the literature. Since zeolite CHA is a microporous structure with a maximum included diameter of 7.37 Å, adsorption of more than one layer of H_2S molecules is unlikely.[11] Therefore, it is expected that Langmuir isotherms should be applicable, which is also reflected on the fitting. Table 2.7 again demonstrates the adsorption predicted using our proposed models agrees well with the experimental data. These results suggest that all of our models developed herein may be used to accurately predict the adsorption of H_2S in all-silica zeolites, thus allowing ones to carry out a large-scale computational study of zeolite adsorbents for the H_2S removal.

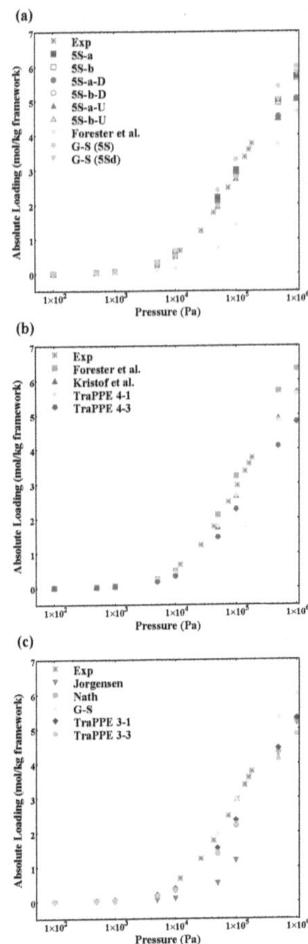

Figure 2.10 Adsorption isotherms of H$_2$S in all-silica zeolite CHA computed using available (a) 5-sites, (b) 4-sites, and (c) 3-sites models, including those developed in this work and reported in the literatures.[1-6] Experimental results from Maghsoudi *et al.* are shown for comparison.[8] For brevity, G-S refers to Gutiérrez-Sevillano *et al.*

Table 2.7 Values of Henry coefficient (K_H) and saturation loading (Q_{sat}) obtained from fitting the adsorption isotherms predicted using the (a) 3-sites, (b) 4-sites, and (c) 5-sites models from this work and those models reported in the literature to the Langmuir equation.[33,34,36,38–40] Fitted values for the experimentally measured isotherm by Maghsoudi *et al.* are also included for comparison.[62]

	$K_H \times 10^5$ (mol/kg/Pa)	Q_{sat} (mol/kg)
Exp[62]	5.95	5.72
(a) 3-sites models		
Jorgensen[36]	1.61	7.96
Nath[40]	3.42	6.05
Gutiérrez-Sevillano *et al.*[33]	5.41	6.73
Shah et al. (TraPPE 3-1)[34]	3.91	6.00
Shah et al. (TraPPE 3-3)[34]	3.76	5.46
(a) 4-sites models		
Forester *et al.*[38]	5.91	7.08
Kristóf and Liszi [39]	4.60	6.36
Shah *et al.* (TraPPE 4-1)[34]	4.75	6.25
Shah *et al.* (TraPPE 4-3)[34]	3.89	5.39
(b) 5-sites models		
5S-a	6.29	6.15
5S-a-D	6.13	5.44
5S-a-U	5.67	5.47
5S-b	6.02	6.10
5S-b-D	6.26	5.41
5S-b-U	5.66	5.45
Gutiérrez-Sevillano *et al.* (5S)[33]	7.19	6.46
Gutiérrez-Sevillano *et al.* (5Sd)[33]	5.16	6.43
Shah *et al.* (TraPPE 4-3)[34]	3.89	5.39

Some of the previously reported models are instead unable to reproduce the experimental adsorption isotherms with accuracy. Predictions by 3-sites models show a notably underestimation in K_H by as much as 4-fold (e.g., the model by Jorgensen, see Table 2.7), in agreement with what is found in Mg-MOF-74. Addition of one more pseudo-sites in the model (i.e., 4-sites) does not show significant improvements. Although the K_H value by the 4-sites model of Forester et al. is closer to experimental data, its predicted Q_{sat} appears to be highly overestimated (~35 %). Interestingly, as opposed to our models, other 5-sites models reported in the literature are also unable to explain the adsorption phenomena of H_2S in all-silica zeolite CHA. To be specific, the 5-sites model proposed by Forester et al. predicts a Henry coefficient value of 1.80×10^{-5} mol/kg/Pa, which is much lower than the experimental one of 5.95×10^{-5} mol/kg/Pa, rather comparable to that predicted by some of the 3-sites models. Given the model by Forester et al. was not developed to optimize its charge distribution to ESP, this may indicate again that achieving a good ESP surface representation is indeed essentially important even though the electrostatic potential is generally regarded to be relatively less important in siliceous zeolites as compared to that in open-metal site MOFs. In summary, these comparisons evidently show that the models developed using our methodology result in accurate and consistent predictions, which also suggests robustness of the methodology.

Model Validation – Liquid Properties

Although a great success has been demonstrated for the H_2S models developed in this work using our methodology for describing adsorbed phase, it remains important to also investigate the behavior of these models for predicting other bulk phase properties in additional to VLE. For this reason, molecular dynamics simulations to compute a variety of bulk liquid properties, including liquid density, radial distribution function (RDF), and self-diffusivity are carried out.

Among those properties, liquid density and RDF are computed using molecular dynamics simulations in an isothermal-isobaric ensemble at 208 K and 1 atm, and the results are summarized in Figure 2.11, with experimental references.[57,63] As shown in the insets of the figure, all of our 5-sites models can accurately reproduce the experimental liquid density. Further, Figure 2.11 shows that, for the RDF between S (H_2S) and H (H_2S), 5-sites models without hydrogen vdW interactions (i.e., 5S-a and 5S-b) are found to predict the experimental S-H RDF precisely. Interestingly, the S-H RDF predictions using models having hydrogen parameters (i.e., 5S-a-U, 5S-b-U, 5S-a-D, and 5S-b-D) show a distinct behavior from the experiment. Specifically, their first peak of the S-H RDF is shifted inward with a more pronounced peak height compared to that from 5S-a and 5S-b or experimental reference. This suggests that models with or without LJ parameters for hydrogen atoms predict different preferential orientations between H_2S molecules in the proximity, and the assigned parameters for hydrogen from DREIDING or UFF may lead to less accurate H_2S local orientations under this condition. On the other hand, as expected from the agreement between calculations and experiments for the density, Figure 2.12 shows that the RDF profiles between S (H_2S) and S (H_2S) predicted by all of our models are similar. Specifically, all models predict the location of the first peak of the S-S RDF at approximately 4 Å, although there are minor variations in the peak height. These results indicate that even though the predictions of bulk liquid properties such as density are approximately the same regardless of whether the hydrogen LJ parameters are considered for a H_2S model, the local orientation can be very dependent on the presence of the hydrogen LJ parameters. There might exist a complicated interplay between the LJ potential parameters and the relative orientation and local distribution of molecules. Therefore, including available RDF in additional to the VLE as a secondary reference for the optimization of LJ parameters can be an important future study, particularly for models that would be employed for studying various micro-level liquid properties, such as their dimer structures. This further parameterization, however, is not our current focus since our developed models utilizing only VLE for LJ parameterization have already achieved satisfactory results for a variety of properties in both adsorbed and bulk liquid phases, as shown by all of the above validations and the self-diffusivity of liquid H_2S discussed below. Table 2.8 summarizes the self-diffusivity of liquid H_2S computed in a microcanonical ensemble at 243.1 K, using 5S-a and 5S-b models. The three-dimensional Einstein's equation is utilized to correlate the translational self-diffusion coefficient and mean square displacement obtained from the molecular simulations. Furthermore, the obtained diffusion coefficients are corrected in order to account for the simulation system-size dependence of the diffusivity.[64] Although our model development methodology is intended for gas molecular models and involves gas-phase DFT calculations in the model development, our simulation results show that these models can also reasonably well reproduce the experimental liquid-phase self-diffusivity by Dupré *et al.*[65] Particularly, the difference of the diffusivity computed with the 5S-a model and by experimental measurements is within 10%.

Overall, these results for the bulk liquid phase density, RDF, and self-diffusivity suggest that the H$_2$S models proposed in this work could be employed not only for studying the adsorptive properties of H$_2$S but also for their bulk liquid properties, with a wide applicability.

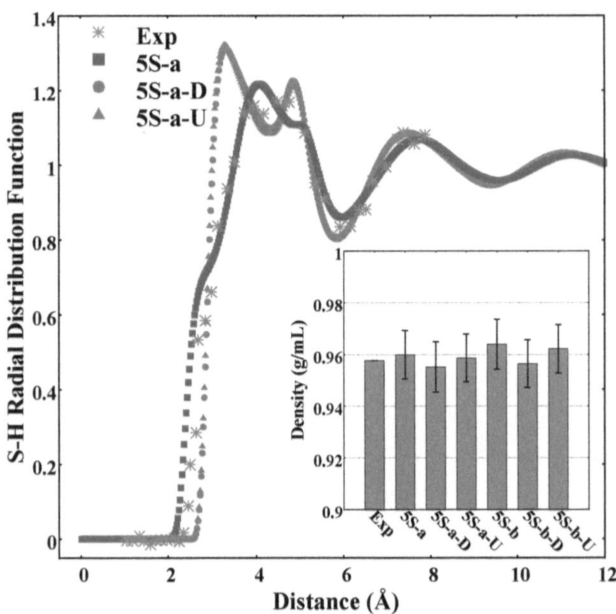

Figure 2.11 Radial distribution function (RDF) between S (H_2S) and H (H_2S) at 208 K and 1 atm using 5-sites models developed in this work. Experimental reference from Andreani et al. is included for comparision.[63] Density predicted with these models at the same condition are presented in the inset with the experimental reference from NIST.[57] In the figures, RDF predictions using 5S-b, 5S-b-D, and 5S-b-U models are not shown as their results are essentially identical to that of 5S-a, 5S-a-D, and 5S-a-U, respectively.

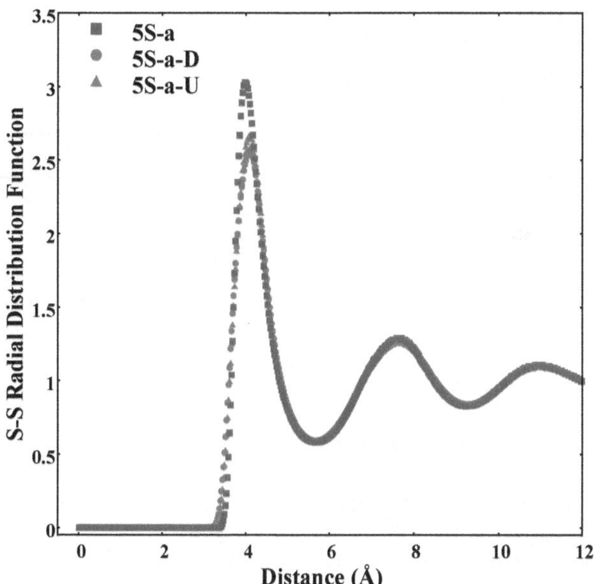

Figure 2.12 Radial distribution function (RDF) between S (H_2S) and S (H_2S) at 208 K and 1 atm using 5-sites models developed in this work. In the figures, RDF predictions using 5S-b, 5S-b-D, and 5S-b-U models are not shown as their results are essentially identical to that of 5S-a, 5S-a-D, and 5S-a-U, respectively.

Table 2.8 Self-diffusivity (D_S) and system-size effect corrected self-diffusivity ($D_{S,bulk}$) of liquid H_2S at 243.1 K predicted using 5S-a and 5S-b models. Experimental data by Dupré *et al.* is included for comparison.[65]

	$D_S \times 10^5$ (cm^2/s)	$D_{S,bulk} \times 10^5$ (cm^2/s)
Exp	-	6.60 ± 0.30
5S-a	5.35	6.14
5S-b	4.99	5.78

2.1.4.3 Computational Screening of All-silica Zeolites

As our models are shown to accurately predict adsorption properties in porous materials, these models can be utilized to efficiently seek for promising materials that are able to selectively capture H_2S from gas mixtures. As a proof of concept, we herein carry out a computational screening, employing the 5S-b model, to explore potential siliceous zeolites. Specifically, more than 160 zeolites included in the IZA database that have been synthesized to date are explored. These zeolites are defect-free (e.g., defective zeolites such as CLO and CHI are excluded) and are geometrically accessible to H_2S. In evaluating the potential of adsorbent materials for capturing molecules of interest, there have been many evaluation criteria considered in many of previous studies. For instance, Snurr and co-workers examined porous materials for carbon capture based on five evaluation criteria, including CO_2 uptake, working capacity, regenerability, selectivity, and selection parameter.[8] Considering that the Henry coefficient of adsorption is highly relevant to many of the five criteria, such as H_2S uptake, working capacity, or selectivity, the Henry coefficient of pure H_2S in all-silica zeolites is focused herein. Among structures studied in this work, 10 zeolites that show the highest Henry coefficient are listed in Table 2.9, together with the computed Henry coefficient (K_H) and isosteric heats of adsorption (Q_{st}), as well as their geometrical and topological features. These structures are predicted to possess Henry coefficients of as large as 4×10^{-3} mol/kg/Pa. These zeolites that show high K_H also have large Q_{st}, indicating that K_H and Q_{st} are strongly related. Indeed, as shown in Figure 2.13 (a), we find a linear correlation between K_H and Q_{st}. For instance, zeolite UEI that was predicted to offer the highest K_H also has the largest isosteric heats of adsorption of nearly 50 kJ/mol. However, there are some structures, such as ABW, AEN, or WEI, which are exceptional to the linearity and have relatively lower K_H at a given Q_{st}. This may be attributed to their very narrow pore or channel width (i.e., D_i (maximum diameter of sphere that can be included) ~ 4.3 Å and D_f (maximum diameter of a sphere that can diffuse through the structure) ~ 3.6 Å). Another noticeable relationship of K_H and topological features is that almost all zeolites with high K_H have their D_i or D_f in the range of 5 ~ 7 Å. This can also be clearly seen in the K_H vs. D_i graph, shown in Figure 2.13 (b). This relationship can serve as an important guideline for the rational design of porous materials, suggesting that zeolites with optimum pore or channel size of 5 ~ 7 Å in diameter have great potential in capturing H_2S.

Table 2.9 Ranking of zeolites based on the predicted Henry coefficient (K_H ($\times$ 10^4 mol/kg/Pa)) of H_2S at 298 K. The corresponding isosteric heats of adsorption (Q_{st} (kJ/mol)) and geometrical and topological features are also shown. D_i (Å): maximum diameter of a sphere that can be included in the structure, D_f (Å): maximum diameter of a sphere that can diffuse through the structure, AV (%) and AA (m^2/g): accessible volume and area, respectively, probed by a sphere of radius 1.4 Å.

Zeolite	K_H	Q_{st}	D_i	D_f	AV	AA
UEI	39.5	48.5	5.60	3.77	6.53	1019
DFT	27.2	43.2	5.10	3.65	6.58	1167
DAC	18.4	43.3	5.28	4.19	9.50	1092
MOR	15.3	43.4	6.70	6.45	12.3	1010
LOV	14.9	43.4	5.15	3.78	7.57	1259
EON	10.5	43.9	7.83	6.79	12.5	1039
RHO	9.26	41.1	10.4	4.06	20.6	1484
RSN	8.08	43.5	5.15	3.77	7.03	1304
MER	6.42	41.0	6.65	4.20	10.4	1376
AWO	5.86	38.3	5.16	3.67	5.35	825.2

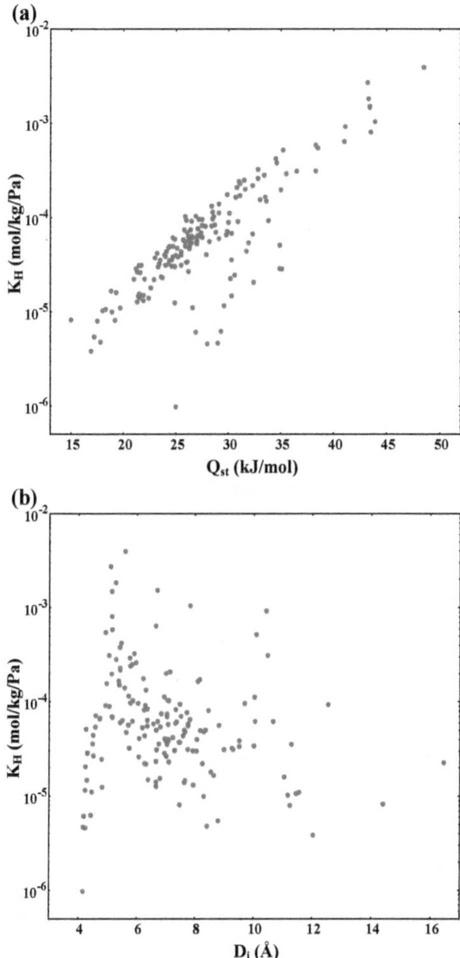

Figure 2.13 Relationship between Henry coefficient (K_H) of H_2S and (a) isosteric heats of adsorption (Q_{st}) of H_2S or (b) the maximum diameter of sphere (D_i) that can be included in

the structure. Henry coefficients and isosteric heats of adsorption are computed using the 5S-b H_2S model developed in the work.

2.1.5 Conclusion

In this work, with the goal of facilitating the discovery of promising nanoporous materials as adsorbents for gas separations, we introduced a systematic, robust, and computationally efficient methodology for the accurate model development of small gas molecules. To adequately describe the electrostatic potential surface surrounding the gas molecules, quantum chemical DFT calculations are incorporated in our approach with the REPEAT scheme to optimize the number and location of massless sites and the overall charge distributions. Subsequently, the van der Waals interaction parameters are fitted to reproduce the experimental vapor-liquid equilibrium. With this method, thanks to the decoupling of the force field parameters, all possible model parameters can be explored in the development procedure, while significant savings in computational cost can be achieved.

In light of the importance of H_2S removal, we applied the methodology for the H_2S model development. Our results suggest that at least two massless sites or more are needed in order to accurately describe the ESP surface, and 5-sites models seem to reach a good balanced point between the accuracy of force field and the required computational cost when being employed in molecular simulations. We clearly demonstrate that the models developed herein show significant improvements in predictions of H_2S adsorption in an all-silica zeolite and metal-organic frameworks, as opposed to other currently available models in the literature. It is found that our models can capture a variety of liquid-phase properties as well. Although H_2S models with high accuracy have been obtained, there can be additional efforts of including the vdW interaction for the hydrogen atoms in the model. Overall, this methodology is shown to provide a systematic, efficient, and robust means toward the development of accurate molecular models for small gases. These accurate H_2S models developed in this work can also facilitate the future computational discovery of potential nanoporous adsorbents to mitigate the toxic gas release. As an example, we have employed these models to identify promising zeolites from more than 160 candidates for the removal of H_2S.

2.2 Electrostatic Potential Optimized Molecular Models (ESP-MMs) for Molecular Simulations: CO, CO_2, COS, H_2S, N_2, N_2O, and SO_2

Reprinted in part with permission from E. H. Cho and L.-C. Lin, *J. Chem. Theory Comput.* 2019, 15, 11, 6323–6332. Copyright © 2020 American Chemical Society

2.2.1 Introduction

We proposed above a systematic and robust methodology for molecular model development that ensures accuracy in the charge distributions of the models.[31,66] This was achieved by systematically optimizing the number of massless sites included in the molecular model with their location and the charge assignments to reproduce the electrostatic potential (ESP) surrounding the molecule computed with accurate QM methods. As a proof of concept, we applied the methodology to hydrogen sulfide (H_2S) molecular models and showed that the developed models have improved accuracy in reproducing both adsorption and bulk liquid properties. Given the success, we herein extend this methodology to several other molecules, including carbon monoxide (CO), carbon dioxide (CO_2), carbonyl sulfide (COS), nitrogen (N_2), nitrous oxide (N_2O), and sulfur dioxide (SO_2). The removal or storage of these molecules are critically important in energy- or environmentally-related processes. For instance, CO is formed from an incomplete combustion of carbon-containing materials,[67] and its removal is essential as the inhalation of CO may cause diseases such as those related to cardiovascular.[68] Acidic molecules such as CO_2 or SO_2 are also known for their negative effects on the environment.[69] It should be noted that, different from the originally proposed methodology, two global scaling factors are introduced in the parameterization of dispersion parameters for molecules with heterogeneous atoms, and this updated approach is also employed to re-parameterize the molecular models for H_2S. Overall, their geometry is first optimized with density functional theory (DFT) calculations and the ESP surrounding the molecule is simultaneously obtained. Subsequently, the number of massless sites, the location of massless sites, and the point charge assignments are optimized by reproducing the DFT-computed ESP. Lastly, with an accurate ESP representation, the vdWs interaction parameters are optimized to fit to the VLE curve. For molecules with heterogeneous atom types (e.g., CO) that can have four or more vdWs interaction parameters, we introduce global scaling factors, one for the epsilon and the other one for the sigma of the Lennard-Jones potential, to significantly reduce the parameter space.

In the followings, we first summarize our methodology. In the subsequent results and discussions section, we introduce the developed molecular models, denoted as electrostatic potential optimized molecular models (ESP-MMs), for all of the abovementioned molecules. The developed CO model is then discussed in detail as a case study, and the importance of the number and location of the massless site is thoroughly explored. We show that the CO model with reliable charge distributions can accurately identify the binding mechanism of CO adsorbed in open-metal site MOF-74 materials and Cu-BTC. We then take a step forward to demonstrate the preferential importance of the ESP by showing that extreme values of Lennard-Jones parameters do not affect the binding orientation provided an accurate ESP representation. The experimental adsorption isotherm of CO in Cu-BTC and all-silica MFI can also be reasonably reproduced with the ESP-MM.

Comparisons with previously developed CO models are included as well. For the rest molecular models, we verify their accuracy in predicting their binding geometry in MOF-74. The adsorption isotherm of SO_2 in all-silica MFI is also computed and compared with experimental measurements. Lastly, we demonstrate the high fidelity of the charge distributions described by these ESP-MMs as their charges are determined to remain nearly identical when molecules are clustered in a dimer configuration.

2.2.2 Methodology

The procedure for the molecular model development involves three steps: (1) geometry and ESP surface, (2) number and location of massless sites and charge assignments, and (3) Lennard-Jones parameters. We follow the same procedure for the first two steps as illustrated above in Section 2.1. Details regarding step 3 can be found below.

2.2.2.1 Lennard-Jones Parameters

The Lennard-Jones (L-J) interaction potential is adopted in these ESP-MMs models for the description of repulsion and dispersion as it is known to possess moderate accuracy and excellent computational efficiency. Although there are other non-polarizable or polarizable potential forms proposed to potentially offer a higher accuracy, L-J potentials are anticipated to be a preferred option for large-scale computational screening studies and are compatible with most (if not all) available simulations packages for broader applicability. The L-J parameters are optimized by fitting to the experimental VLE curve. For molecular models that have heterogeneous atom types with four or more L-J parameters, to reduce the parameter space, we introduce in this study two global scaling factors; one for epsilon (S_ε) and the other one for sigma (S_σ) as shown in Equation 2.4 below.

$$E_{L-J} = 4S_\varepsilon\varepsilon\left[\left(\frac{S_\sigma\sigma}{r}\right)^{12} - \left(\frac{S_\sigma\sigma}{r}\right)^{6}\right] \qquad (2.4)$$

Based on well-established generic force fields Dreiding[51], the two scaling factors are applied to scale the parameters from Dreiding and optimized to reproduce experimental VLE. For each combination of the scaling factors, VLE data over an extensive temperature range are computed by the Gibbs ensemble Monte Carlo (GEMC) technique. The one with the least error as compared to the experimental values are chosen as the final L-J parameters. The corresponding critical temperature and density are also calculated using the scaling law[45] and the law of rectilinear diameters[46], respectively, and compared with experimental data.

With the molecular models developed from the previous steps, a variety of adsorption properties can be investigated for validations. More computational details can be found below.

2.2.3 Computational Details

2.2.3.1 Density functional theory (DFT) calculations

Geometry optimization and electrostatic potential (ESP) calculations can be found in Section 2.1. Details regarding interaction energy calculations can be seen in the followings.

Interaction Energy Calculations: DFT calculations are also employed in this study to compute the interaction energies between adsorbed molecules and the host framework.

Specifically, self-consistent field calculations are conducted to compute the interaction energy of a carbon monoxide (CO) molecule adsorbed in a host framework using the Quantum ESPRESSO[43] implementation. A fully optimized structure of Mg-MOF-74 as well as the binding geometry of CO in the structure are obtained from a previous study by Lee *et al.*[27] The most energetically favorable configuration is where the distance between the carbon atom of the CO molecule and the magnesium atom of the framework is 2.6 Å (see Figure 2.27 (a)). With the reported binding configuration at our disposal, other configurations are also probed. Specifically, a flipped configuration of CO (i.e., carbon and oxygen of the CO molecule are switched) are also studied. For these configurations, the energy is computed with Perdew-Burke-Ernzerhof (PBE) exchange-correlation functional with a dispersion-corrected van der Waals functional (vdW-DF2).[70] The kinetic cut off is set to 100 Rydberg, while a 2×2×2 k-point sampling is utilized. The interaction energies are calculated as

$$\Delta E = E_{MOF+mol} - E_{MOF} - E_{mol} \qquad (2.5)$$

where $E_{MOF+mol}$, E_{MOF}, and E_{mol} are the energy of the molecule adsorbed in MOF, bare MOF, and molecule in the gas phase, respectively.

2.2.3.2 Monte Carlo Simulations

Computational details for computing vapor liquid equilibrium (VLE) and critical properties, adsorption isotherms, and binding geometry and energy can be found above in Section 2.1. Below are details regarding energy landscapes and structure and force field information. Monte Carlo simulation are conducted in the RASPA package[44].

Binding geometry and energies: Binding geometry and energy of a given molecular model in nanoporous materials are computed using MC simulations in a canonical ensemble (NVT simulations). A single molecule is placed in the simulation box (i.e., structural framework). Multiple replica of NVT simulations are carried out at a low temperature of 1 K. The configuration with the lowest energy amongst all configurations is identified as the binding geometry.

Energy landscape: Interaction energy between CO and Mg-MOF-74 at varied configurations of CO are probed. Specifically, with an initial binding configuration of CO in Mg-MOF-74 predicted with DFT by Lee et al.[27], various configurations are generated by leaning the CO atom toward one of the oxygen atom of the open-metal site of Mg-MOF-74. Configurations are generated for different angles between the normal vector generated by the four oxygen atoms of the open-metal site and the CO vector. A schematic illustration can be seen in Figure 2.14. The different types of oxygens are specified according to Figure 2.15 (a).

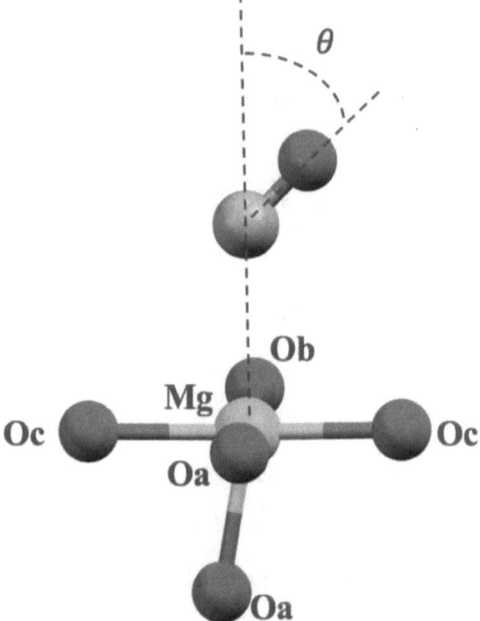

Figure 2.14 Schematic illustration of configurations generated to probe the energy landscape of CO adsorbed near the metal site of Mg-MOF-74. Configurations are generated with different angles (i.e., θ) between the normal vector generated by the four oxygen atoms of the open-metal site and the CO vector.

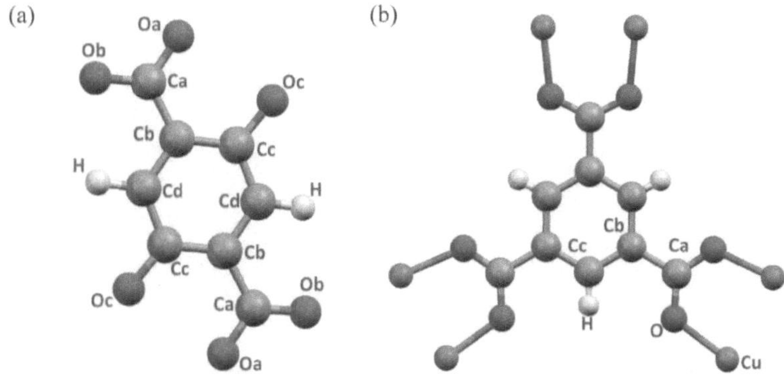

Figure 2.15 Definition of the linker atom types in (a) MOF-74 series and (b) Cu-BTC.

Structures and force field parameters: The crystallographic structure data of zeolites is taken from the International Zeolite Association (IZA).[11] The force field parameters for the zeolite framework are adopted from CLAYFF.[48] For M-MOF-74 (M = Mg, Mn, Fe, Co, Ni, and Zn), the DFT-optimized structure and DFT-derived charges from previous studies are used.[41,61] The L-J parameters of the framework atoms are assigned from the universal force field (UFF).[49] For Cu-BTC, an experimental structure reported from Chui et al. is utilized.[71] The charges of the framework atoms are obtained from a previous computational study by Castillo et al.[72] For the L-J parameters, Dreiding[51] force field is obtained, except the Cu from UFF[49]. The Lorentz-Berthelot mixing rule is applied to estimate the L-J parameters between dis-similar atoms. Void fraction was computed by Widom particle insertion technique with helium as the probe molecule.[30] Helium model is adopted from Hirschfelder et al.[73] In all of the simulations carried out in this study, L-J potential is truncated and shifted to zero at a cutoff radius of 12 Å, while Coulombic interactions are computed using the Ewald summation technique with a relative error of 10^{-6}. The force field parameters are summarized in Table 2.10.

Table 2.10 Point charges and Lennard-Jones (L-J) parameters assigned to (a) all-silica MFI, (b) Mg-MOF-74, (c) Mn-MOF-74, (d) Fe-MOF-74, (e) Co-MOF-74, (f) Ni-MOF-74, (g) Zn-MOF-74, and (h) Cu-BTC. CLAYFF are adopted for all-silica MFI.[48] The charges for MOF-74 series are from previous studies[41,61], whereas UFF is employed for their L-J parameters.[49] For Cu-BTC, charges are taken from Castillo et al.[72] while the L-J parameters are obtained from Dreiding[51] except Cu from UFF.[49] Definition of the linker atom types in MOF-74 series and Cu-BTC are provided in Figure 2.15.

(a)

atom	Point Charge (e^-)	Lennard-Jones (ε (K), σ (Å))
Si	2.10	0.000926, 3.302027
O	-1.05	78.20032, 3.165541

(b)

atom	Point Charge (e^-)	Lennard-Jones (ε (K), σ (Å))
Mg	1.560	55.86 2.69

Oa	-0.899	30.19 3.12
Ob	-0.752	30.19 3.12
Oc	-0.903	30.19 3.12
Ca	0.900	52.84 3.43
Cb	-0.314	52.84 3.43
Cc	0.456	52.84 3.43
Cd	-0.234	52.84 3.43
H	0.186	22.14 2.57

(c)

atom	Point Charge (e^-)	Lennard-Jones (ε (K), σ (Å))
Mn	1.343	9.717 2.638
Oa	-0.754	30.19 3.12
Ob	-0.717	30.19 3.12
Oc	-0.806	30.19 3.12
Ca	0.850	52.84 3.43
Cb	-0.296	52.84 3.43
Cc	0.396	52.84 3.43
Cd	-0.203	52.84 3.43
H	0.187	22.14 2.57

(d)

atom	Point Charge (e^-)	Lennard-Jones (ε (K), σ (Å))
Fe	1.288	6.542 2.594
Oa	-0.753	30.19 3.12
Ob	-0.707	30.19 3.12
Oc	-0.794	30.19 3.12
Ca	0.870	52.84 3.43

Cb	-0.337	52.84 3.43
Cc	0.432	52.84 3.43
Cd	-0.195	52.84 3.43
H	0.196	22.14 2.57

(e)

atom	Point Charge (e^-)	Lennard-Jones (ε (K), σ (Å))
Co	1.189	7.045 2.559
Oa	-0.720	30.19 3.12
Ob	-0.673	30.19 3.12
Oc	-0.725	30.19 3.12
Ca	0.846	52.84 3.43
Cb	-0.308	52.84 3.43
Cc	0.391	52.84 3.43
Cd	-0.177	52.84 3.43
H	0.177	22.14 2.57

(f)

atom	Point Charge (e^-)	Lennard-Jones (ε (K), σ (Å))
Ni	1.298	7.548 2.525
Oa	-0.789	30.19 3.12
Ob	-0.696	30.19 3.12
Oc	-0.785	30.19 3.12
Ca	0.895	52.84 3.43
Cb	-0.349	52.84 3.43
Cc	0.418	52.84 3.43
Cd	-0.173	52.84 3.43
H	0.181	22.14 2.57

(g)

atom	Point Charge (e⁻)	Lennard-Jones (ε (K), σ (Å))
Zn	1.209	62.4 2.46
Oa	-0.719	30.19 3.12
Ob	-0.671	30.19 3.12
Oc	-0.740	30.19 3.12
Ca	0.841	52.84 3.43
Cb	-0.298	52.84 3.43
Cc	0.376	52.84 3.43
Cd	-0.170	52.84 3.43
H	0.172	22.14 2.57

(h)

atom	Point Charge (e⁻)	Lennard-Jones (ε (K), σ (Å))
Cu	1.248	2.52 3.11
O	-0.624	48.2 3.03
Ca	0.494	47.9 3.47
Cb	0.130	47.9 3.47
Cc	-0.156	47.9 3.47
H	0.156	7.65 2.85

2.2.4 Results and Discussion

With the extended methodology as described above, we develop molecular models for CO, CO_2, COS, H_2S, N_2, N_2O, and SO_2. The geometry of these ESP optimized molecular models (ESP-MMs) with the optimal location of the massless sites if applicable are shown in

Figure 2.16. The corresponding point charges and Lennard-Jones (L-J) parameters are summarized in Table 2.11 and Table 2.12, respectively. For some of these models, massless sites are required in order to achieve a better representation of the ESP surrounding the molecule (e.g., CO, COS, N_2, and SO_2 possess one massless site and H_2S two massless sites). We note that the location of massless sites and the charge assignment for H_2S are adopted from our study directly (i.e., 5S-a model, see section 2.1 above), but with the L-J parameters being re-parameterized. The RRMS value of each optimized model can be also seen in Table 2.11. For each ESP-MM developed herein, the RRMS value is lower than 20 %, an indicative of a sufficiently accurate ESP representation. This study proposes for the very first time, to the best of our knowledge, a collection of molecular models that are developed to ensure accuracy in reproducing the ESP representation and the VLE curve in a systematic and efficient manner. We note that, while there are some molecular models reported in the literature, such as that of TraPPE CO_2 or N_2 models[74], that have been successfully implemented in many of the molecular simulation studies[53,75–78], we still include these molecules in the development using our proposed methodology with the aim of offering a consistent and complete set of molecule models. Interestingly, for CO_2 and N_2 molecules, the point charges assigned following our methodology are very similar to the TraPPE models (e.g., -0.3253 vs. -0.35 for the oxygen atoms of CO_2).

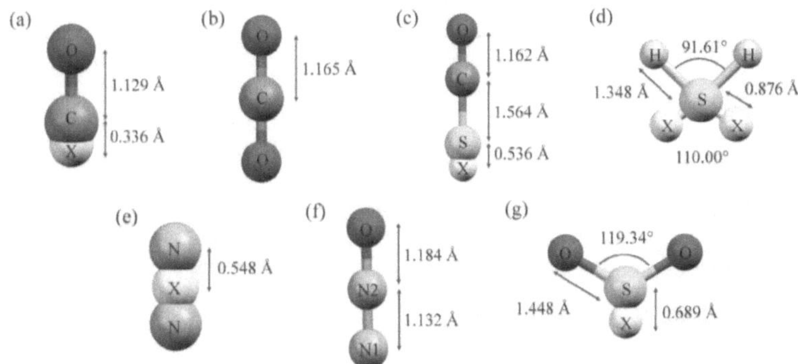

Figure 2.16 Geometry of ESP-MMs developed in this work. (a) carbon monoxide (CO), (b) carbon dioxide (CO_2), (c) carbonyl sulfide (COS), (d) hydrogen sulfide (H_2S) (e) nitrogen (N_2), (f) nitrous oxide (N_2O), and (g) sulfur dioxide (SO_2). Massless sites are specified as X.

Table 2.11 Charge distributions of each ESP-MM developed in this study (see Figure 2.16 for the geometry) and the corresponding RRMS error.

ESP-MM	Point Charge (e^-)				RRMS (%)
CO	O(-0.209)	C(1.040)	X(-0.831)		13.9
CO_2	O(-0.3253)	C(0.6506)			10.0
COS	O(-0.2772)	C(0.5448)	S(-0.7698)	X(0.5022)	19.6
H_2S	S(-1.152)	H(0.268)	X(0.308)		16.3
N_2	N(-0.5186)	X(1.0372)			18.9

| N$_2$O | O(-0.2774) | N2(0.5923) | N1(-0.3149) | | 13.0 |
| SO$_2$ | O(-0.378) | S(1.077) | X(-0.321) | | 9.6 |

Table 2.12 Lennard-Jones (L-J) parameters of each ESP-MM developed in this study (see Figure 2.16 for the geometry). The initial parameters are obtained from Dreiding[51] and are scaled to fit to experimental vapor liquid equilibrium. No parameters are assigned to massless sites.

ESP-MM	L-J parameters (ε (K), σ (Å))		
CO	O(40.5, 3.08)	C(40.2, 3.52)	
CO$_2$	O(56.4, 2.91)	C(56.0, 3.33)	
COS	O(48.7, 3.00)	C(48.4, 3.44)	S(174.8, 3.55)
H$_2$S	S(233.7, 3.48)	H(10.3, 2.76)	
N$_2$	N(38.0, 3.31)		
N$_2$O	O(71.3, 2.91)	N2(57.6, 3.13)	N1(57.6, 3.13)
SO$_2$	O(45.3, 3.00)	S(162.7, 3.55)	

In the followings, we primarily focus on the development of CO model and its validations in detail as an example. To offer comprehensive analysis, we also compare with other available CO models. There have been various molecular models for CO reported in the literature. Sirjoosingh et al. reported a 2-sites model, where the charges are obtained by Mulliken analysis and the L-J parameters from UFF.[79] Piper et al. proposed a 4-sites model with the point charges fitted to reproduce the multipole moments.[80] A 3-sites model was suggested by Straub and Karplus with the massless site assigned to the center-of-the-mass (COM) and the point charges optimized to reproduce the experimental quadrupole moment.[81] Similarly, a 3-sites model, reported by Martín-Calvo et al., has the same molecular geometry and the location of massless site as that of Staub and Karplus,[82] but with the point charges optimized by reproducing the experimental dipole moment and with the L-J parameters optimized by reproducing the experimental VLE curve. It should be noted that the model of Martín-Calvo et al. has the same ratio of the point charges and the L-J parameters among each site as that of the Straub and Karplus. For these models, accurate representation of the ESP surrounding the molecule was undervalued, and not all of the model parameters were systematically included in the optimization.

For the development of the ESP-MM for CO, we start from zero massless site (i.e., 2-sites model), which yields a model with a large RRMS value of 95 %. Per the methodology, an additional massless site is subsequently included (i.e., 3-sites model), which can significantly lower the RRMS value to 13.9%. Interestingly, the optimal massless site is located outside of the C-O bond and near the carbon atom as shown in Figure 2.16 (a). As a comparison, if we fix the location of the massless site at the center-of-the-mass (COM) of the CO molecule, a location that has been widely used in the currently available models as discussed above, the lowest RRMS (i.e., the best possible ESP description) that can be achieved is approximately 30 %. The corresponding model is denoted as the CO(COM) model hereafter. To visualize the ESP representation, Figure 2.17 shows the ESP surface (or the topology of the ESP field) exerted by the CO molecule, computed by DFT and by molecular models.

It can be evidently seen that the ESP by the 2-sites model (Figure 2.17 (b)) significantly deviates from the DFT-computed one (Figure 2.17 (a)), which can be instead well described by our developed 3-sites ESP-MM (Figure 2.17 (c)). Interestingly, while the ESP of the CO(COM) model (Figure 2.17 (d)) seemingly resembles the DFT-computed one reasonably well, assigning the massless site at the COM fails to accurately describe the ESP near the carbon and oxygen atoms of the CO molecule. For a clear presentation, Figures 2.17 (e) and 2.17 (f), respectively, demonstrate the error in the ESP description relative to DFT for the optimized 3-sites model and the CO(COM) model. While minor error is found for our developed 3-sites model, the COM model notably overestimates the ESP near the carbon atom (i.e., too positive) and underestimate the ESP surrounding the oxygen atom (i.e., too negative). We also show the ESP represented by previously developed VLE-fitted CO molecular models in Figure 2.18 with the corresponding RRMS error provided.

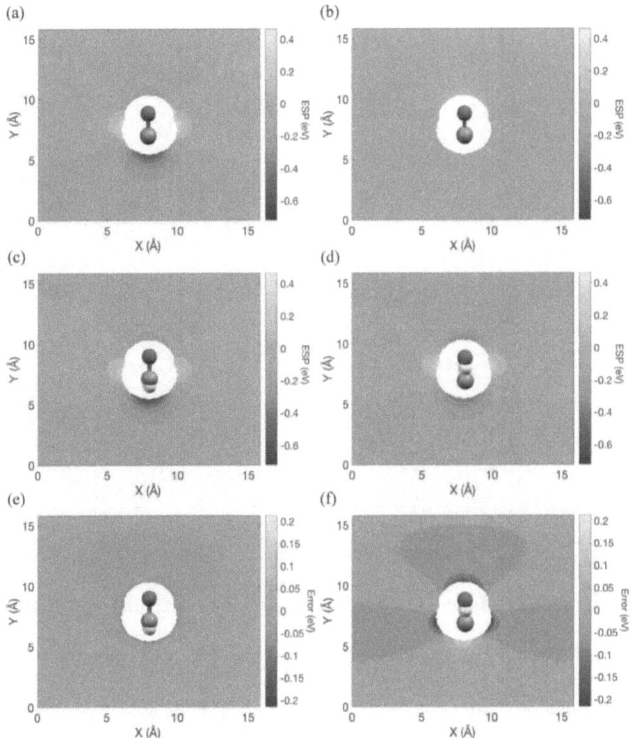

Figure 2.17 Electrostatic potential (ESP) (eV) surrounding the CO molecule computed by (a) DFT and (b-d) point charge methods of (b) 2-sites model, (c) 3-sites model with the optimized location of the massless site (i.e., CO ESP-MM), and (d) 3-sites model with the location of the massless site assigned to be at the center of mass (i.e., CO(COM)). The error in the ESP description of (c) and (d) relative to the DFT-computed one are shown respectively in (c) and (f).

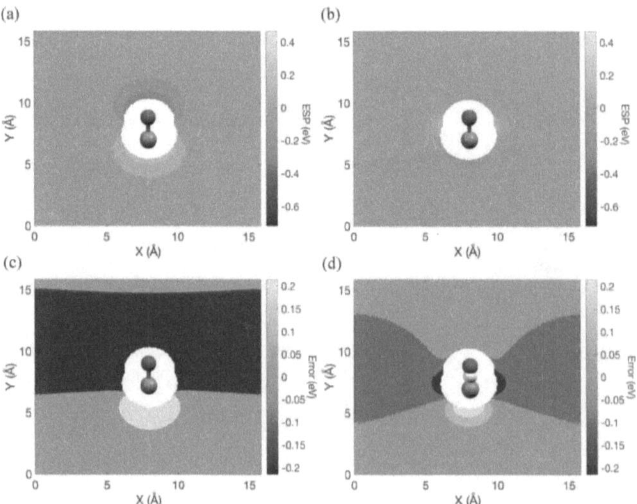

Figure 2.18 Electrostatic potential (ESP) surrounding CO represented by previously developed VLE-fitted CO molecular models by (a) Sirjoosingh et al.[79] (RRMS: 161 %) and (b) Martín-Calvo et al. (RRMS: 62 %)[82] The corresponding deviations in the ESP representation with respect to DFT-computed ESP are shown in (c) and (d), respectively.

With the optimized number and location of massless sites and the corresponding point charges, the L-J parameters of the models are subsequently determined by reproducing the experimental VLE. To reduce the overall parameter space as noted above, based on the assumption that the relative magnitude in the L-J parameters of generic FFs are accurate, two global scaling factors are applied (one for epsilon and the other for sigma) and optimized. Specifically, as noted above, the initial L-J parameters from Dreiding[51] are adopted herein for its popularity in the literature. The VLE curve computed with the CO

ESP-MM can be seen in Figure 2.19, which well reproduces the experimental VLE as well

as the critical temperature and pressure reported from NIST.[57] The L-J parameters for the

CO(COM) model are also optimized (see Figure 2.20 for its VLE curve), and the obtained

parameters as given in Table 2.13 are very similar to that of the CO ESP-MM model. The

same procedures are carried out for the rest of the CO_2, COS, H_2S, N_2, N_2O, and SO_2.

Similarly, Figure 2.21-2.26 show that all these ESP-MMs can resemble experimental VLE

accurately. Interestingly, we find that, by using such a scaling approach, only one optimal

set of scaling factors is identified for each case. By contrast, if all parameters are fitted for

molecules with heterogeneous atoms, multiple optimal L-J parameter combinations are

found and each of them can equally accurately reproduce the VLE curve.

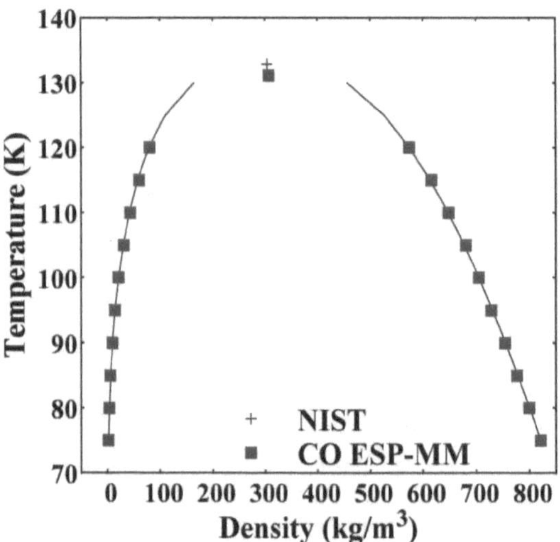

Figure 2.19 Vapor liquid equilibrium (VLE) computed by the CO ESP-MM developed in this work. The standard deviations are smaller than the size of the data points. The experimental reference from NIST is included for comparison.[57]

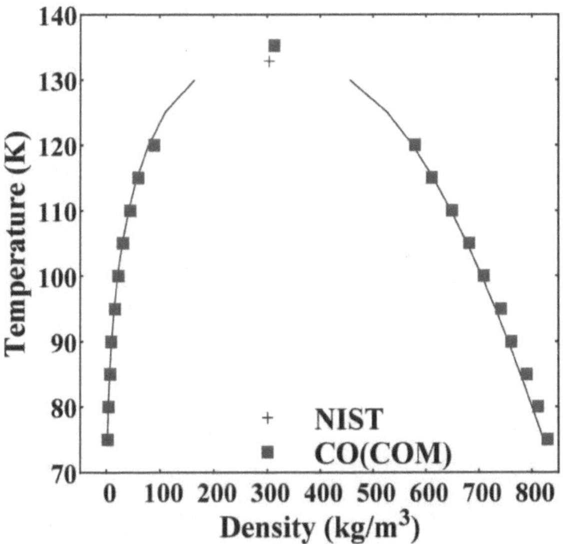

Figure 2.20 Vapor liquid equilibrium (VLE) computed by the CO(COM) developed in this work to compare with CO ESP-MM model. The location of massless site in the CO(COM) model is assigned to be at the center-of-the-mass of the CO molecule instead of being included in the model optimization. The experimental reference from NIST is included for comparison.[57]

Table 2.13 Lennard-Jones (L-J) parameters of three other CO models developed in this study to compare with the CO ESP-MM reported in the main text. For the CO(COM) model, the location of massless site is assigned to be at the center-of-the-mass of the CO molecule instead of being included in the model optimization. The L-J parameters are optimized by scaling the initial L-J parameters from Dreiding.[51] Other two CO models, denoted as CO(U) and CO(M-C), have the same geometry and charge assignments as the CO ESP-MM. These models are developed based on different initial L-J parameters as compared to the CO ESP-MM, which are from UFF[49] and Martín-Calvo et al.[82], respectively.

	Lennard-Jones (atom(ε (K), σ (Å)))	
CO(COM)	O(40.5, 3.06)	C(40.2, 3.50)
CO(U)	O(29.0, 3.15)	C(50.7, 3.46)
CO(M-C)	O(92.1, 3.01)	C(15.2, 3.68)

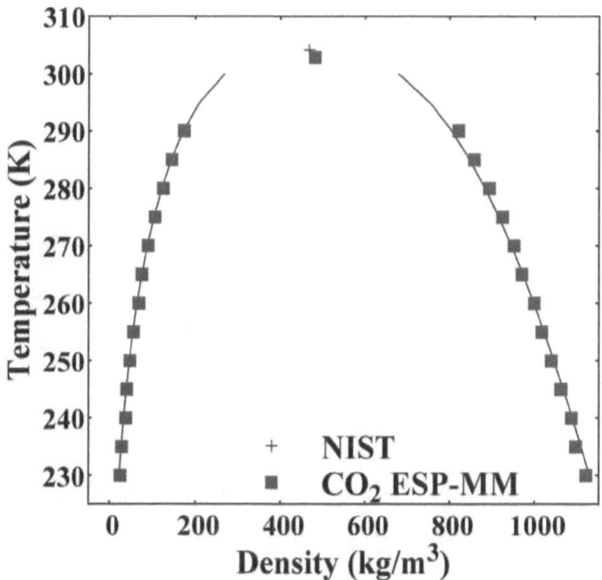

Figure 2.21 Vapor liquid equilibrium (VLE) computed by the CO_2 ESP-MM developed in this work. The experimental reference from NIST is included for comparison.[57]

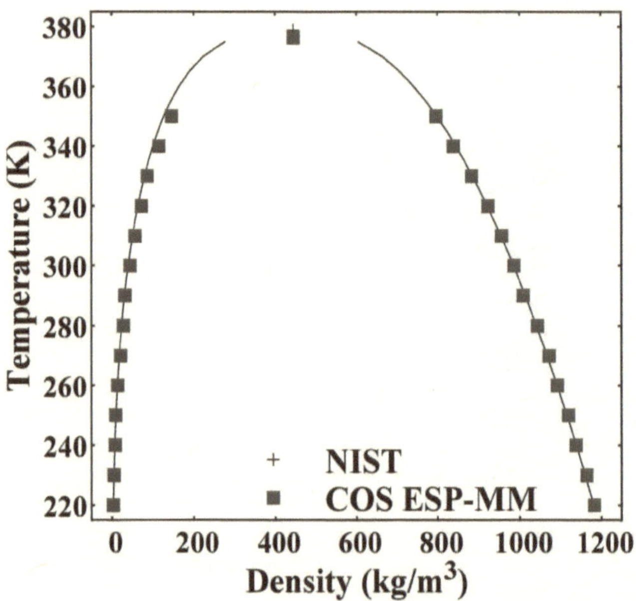

Figure 2.22 Vapor liquid equilibrium (VLE) computed by the COS ESP-MM developed in this work. The experimental reference from NIST is included for comparison.[57]

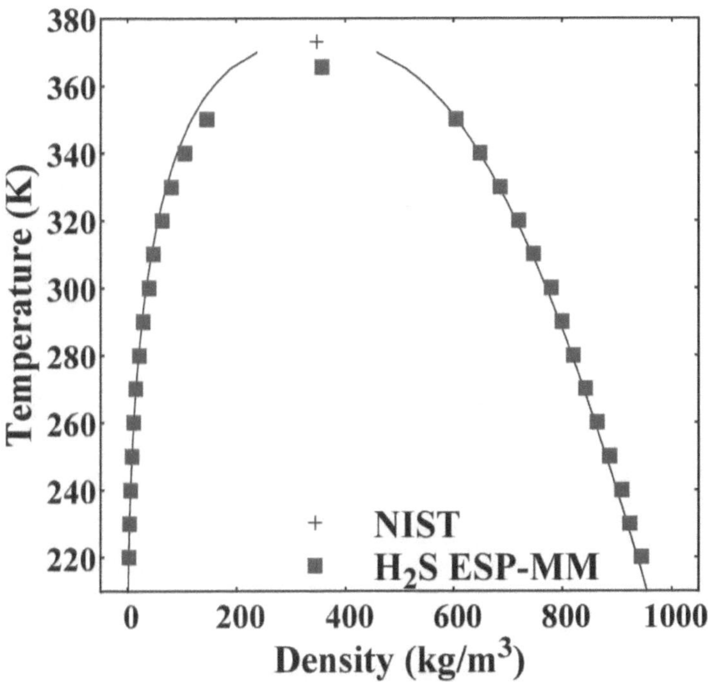

Figure 2.23 Vapor liquid equilibrium (VLE) computed by the H₂S ESP-MM developed in this work. The experimental reference from NIST is included for comparison.[57]

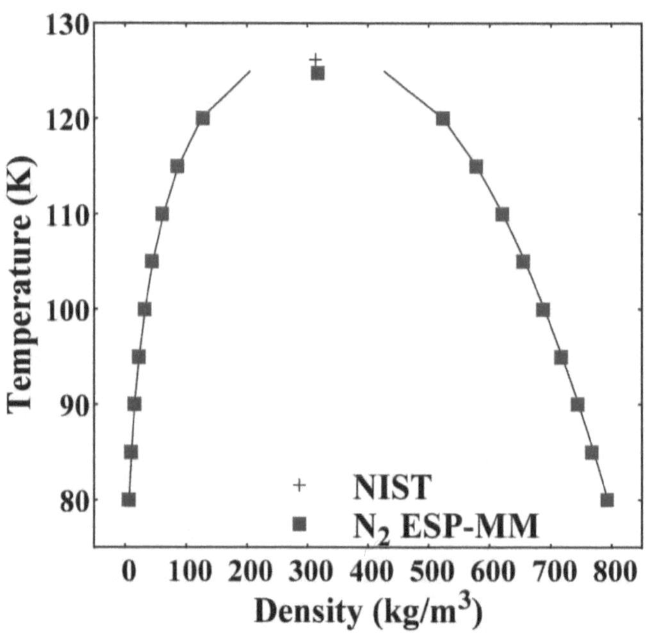

Figure 2.24 Vapor liquid equilibrium (VLE) computed by the N$_2$ ESP-MM developed in this work. The experimental reference from NIST is included for comparison.[57]

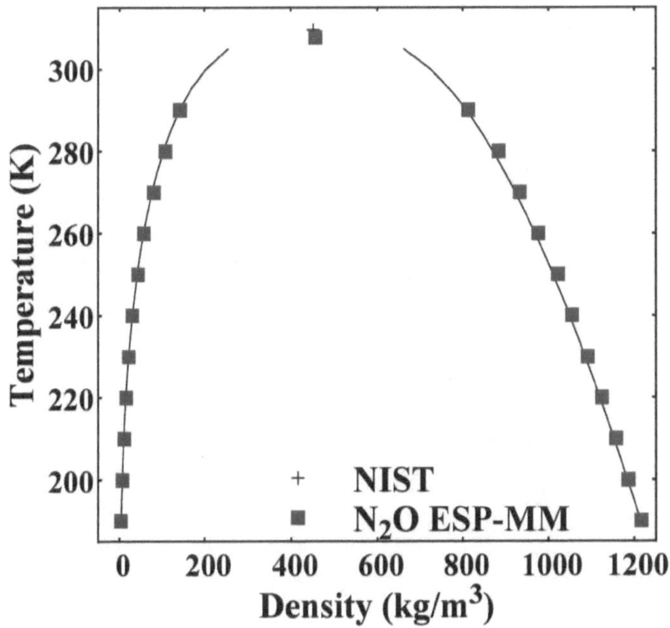

Figure 2.25 Vapor liquid equilibrium (VLE) computed by the N_2O ESP-MM developed in this work. The experimental reference from NIST is included for comparison.[57]

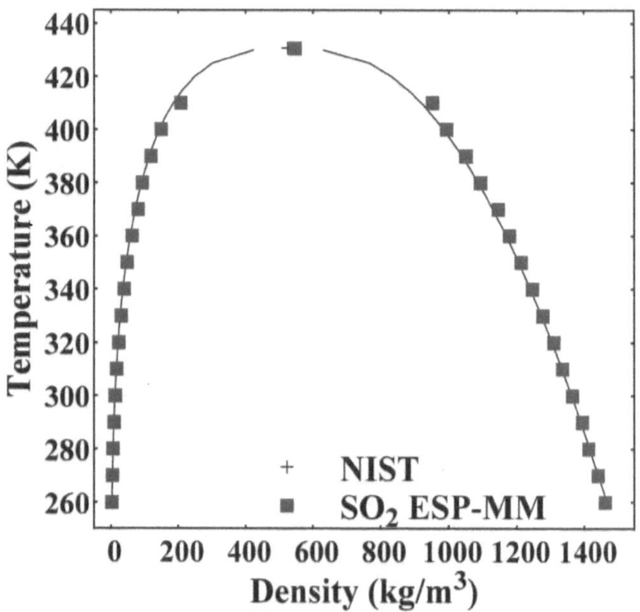

Figure 2.26 Vapor liquid equilibrium (VLE) computed by the SO$_2$ ESP-MM developed in this work. The experimental reference from NIST is included for comparison.[57]

To validate the developed CO ESP-MM, adsorption properties of CO in various nanoporous materials are computed, with specific emphasis on open-metal site structures that provide high affinity towards molecules that possess dipole or quadrupole moments. In these structures, electrostatic interactions play an important role in dominating adsorption properties. First, we compute the binding configuration of CO in Mg-MOF-74, a material that has been extensively studied both experimentally[83,84] and computationally[27].

From previous DFT or experimental studies, CO binds to the Mg-MOF-74 in a way where the carbon atom is closer to the open-metal site than the oxygen atom (i.e., "Mg-C-O" configuration) as shown in Figure 2.27 (a). With our ESP-MM, Figure 2.27 (b) shows that model with accurate ESP are indeed also able to correctly describe the binding mechanism, with a "Mg-C-O" configuration. By contrast, an opposite configuration of CO (i.e., "Mg-O-C" configuration) is predicted with the CO(COM) model or previous CO models reported in the literature (i.e., models by Sirjoosingh et al and Martín-Calvo et al.) as shown in Figure 2.27 (c). Provided that the L-J parameters assigned to the CO ESP-MM and the CO(COM) model are nearly identical, this flipped configuration predicted with the CO(COM) model can be attributed to the insufficient ESP representation. As discussed above and shown in Figure 2.17 (f), CO(COM) model presents over-prediction and under-prediction of the ESP near the carbon atom and the oxygen atom, respectively, as a result of the location of the massless site being designated to the center-of-the-mass rather than being included in the optimization. Although both the CO ESP-MM and the CO(COM) model are developed by reproducing VLE (i.e., configurations occurred in the liquid/gas phase), by including electronic structure calculations in the model development to ensure accurate ESP representations, the ESP-MM can also offer more accurate predictions in the spatial configuration of the adsorbed molecules.

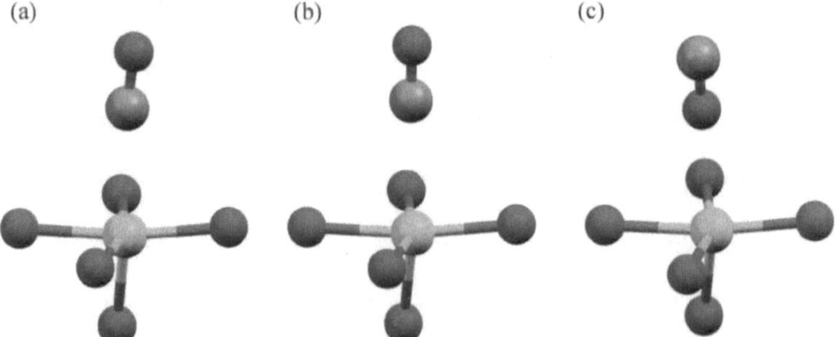

Figure 2.27 Binding geometry of CO adsorbed in Mg-MOF-74 predicted with (a) DFT calculations by Lee et al.[27], (b) the CO ESP-MM, and (c) the CO(COM) model. The binding orientation shown in panels (a) and (b) is denoted as the "Mg-C-O" configuration, whereas that displayed in panel (c) is referred to as the "Mg-O-C" configuration.

We also compute the binding energy using the developed CO ESP-MM. An underestimation in binding energy as compared to that predicted by DFT from Lee et al. is observed (~28 kJ/mol vs. 39 kJ/mol).[27] This is in fact expected as many previous studies reported in the literature have found that the use of L-J potential with parameters from generic FFs for open-metal site structures may lead to too repulsive vdW interactions, which accordingly underestimates the host-guest interaction energy.[32,41] Parametrizing the host-guest interaction for better describing the non-Coulombic contribution of the interactions between the adsorbent and the adsorbate, particularly in open-metal site structures, has therefore been an active research area.[32,41,61,85,86] While parameterizing the adsorbent-adsorbate non-Coulombic interaction is out of the scope of this study, it is important to note that having an accurate molecular model that can ensure accurate description of the Coulombic part such as ESP-MMs is essential in such parametrization. These developed ESP-MMs can ensure that the Coulombic interaction between the framework of interest and the guest molecule can be more accurately described, such that additional corrections will not be included in the fitted dispersion contribution of the parametrized potential. Given the host-guest interactions for open-site MOFs are not specifically optimized, to better validate our models with the DFT-computed energies, relative energies between the two different adsorption states (i.e., "Mg-C-O" and "Mg-O-C") are compared and summarized in Figure 2.28 (a). Figure 2.28 (a) shows our CO ESP-MM can compute the relative energies accurately, whereas the CO(COM) model or previously developed models of CO predict, as anticipated, an opposite binding behavior. To further compare the behavior of CO ESP-MM relative to others, multiple CO adsorption configurations are generated by rotating a carbon monoxide molecule near the open-metal site to probe their description in the energy landscape. Specifically, the angle between the C-O vector of CO molecule and the normal vector of the plane generated with four oxygens connected to the Mg are varied (see Figure 2.14 for a schematic illustration). Figure 2.28 (b) shows the interaction energy between the CO and the adsorbent for these configurations computed with DFT and models of the CO ESP-MM and that by Sirjoosingh et al.[79] and Martín-Calvo et al.[82], and the CO ESP-MM is able to follow the same trend in the energy as that predicted by DFT. While the over predictions in the energy with CO ESP-MM is expected as discussed above, the difference is consistent throughout different configurations.

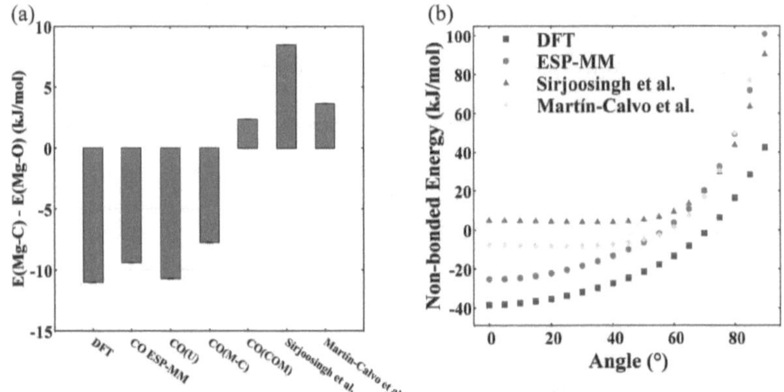

Figure 2.28 (a) Differences in the interaction energy (kJ/mol) between the "Mg-C-O" (Figure 2.27 (b)) and the "Mg-O-C" (Figure 2.27 (c)) configuration of CO adsorbed in Mg-MOF-74. (b) Energy landscape of the interaction of CO rotated near the open-metal site of Mg-MOF-74 (see Figure 2.14). The interaction energies are computed using DFT calculations or classical simulations employing models developed in this study and available CO models in the literature.[79,82]

To investigate the effect of L-J parameters, we also develop CO models with the geometry and point charges same as that of the CO ESP-MM but based on different initial L-J parameters. Specifically, we adopt L-J parameters from UFF[49], another widely used generic FF, and also from the model by Martín-Calvo et al.[82] The optimized L-J parameters are summarized in Table 2.13, where the models are denoted as CO(U) and CO(M-C), respectively. The VLE predicted with these models are shown in Figure 2.29. The L-J parameters from Dreiding and UFF can be deemed comparatively similar to each other. Therefore, as expected, the resulting CO ESP-MM and the CO(U) models yield almost the same predictions, for the binding geometry and energy as well as the relative energy between the "Mg-C-O" and "Mg-O-C" configurations. On the other hand, the parameters used for developing CO(M-C) are drastically different. The epsilon value of the oxygen atom is much greater than that of the carbon atom by 6.1 times, whereas the ratio is within a factor of 1.7 for Dreiding or UFF. Interestingly, even with such a different set of parameters, the CO(M-C) model still predicts a "Mg-C-O" configuration and nearly identical binding energy as that by the CO ESP-MM although it computes a smaller relative energy difference as shown in Figure 2.28 (a). Overall, all of these results again reinforce the importance of developing molecular models with a high accuracy in the ESP representation, but has yet been largely overlooked to date.

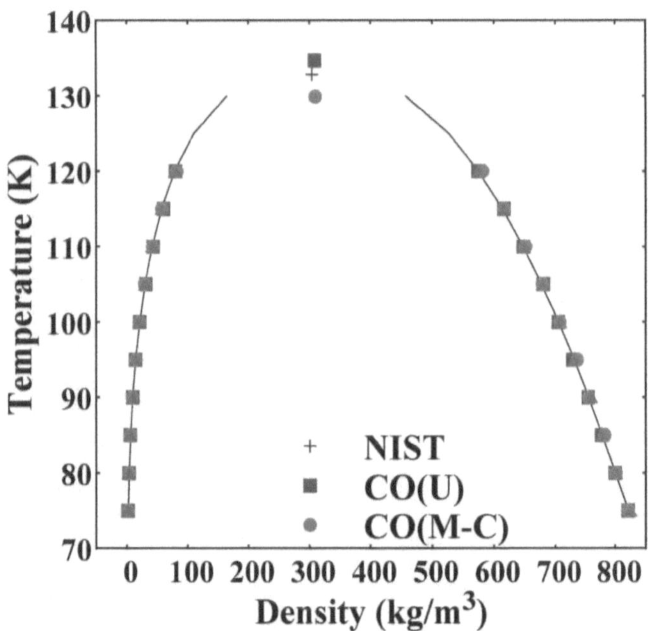

Figure 2.29 Vapor liquid equilibrium (VLE) computed by the CO(U) and CO(M-C) models developed in this work. These models possess same molecular geometry and point charges as that of CO ESP-MM model but have different Lennard-Jones parameters. The Lennard-Jones parameters for these models are summarized in Table 2.13. The experimental reference from NIST is included for comparison.[57]

As further validations for the CO ESP-MM, we also compute the binding configuration and binding energy of CO in M-MOF-74 (i.e., M=Mn, Fe, Co, Ni, Zn), which is isoreticular to the Mg-MOF-74 but with different chemical environment. Furthermore, we also investigate the binding geometry of CO in Cu-BTC, a copper-based paddle wheel open-metal site structure. The binding configurations computed with our model in M-MOF-74 and Cu-BTC are shown in Figure 2.30 and Figure 2.31, respectively. The binding energy calculations in M-MOF-74 are shown in Figure 2.32. Calculations with the model by Sirjoosingh et al.[79] and Martín-Calvo et al.[82] are also included. Overall, our ESP-MM is again able to consistently predict the binding mechanism as reported experimentally[83] and/or computationally[27,87]. Furthermore, we compute the adsorption isotherm of CO in Cu-BTC at 295 K and all-silica MFI zeolite at 298 K, which the experimental measurements by Chowdhury et al.[88] and Matito-Martos et al.[89], respectively, are available and these results can be respectively seen in Figure 2.33 (a) and (b). Our calculations show a reasonable agreement with the experimental adsorption isotherms provided that the adsorbate-adsorbent interaction has not yet been optimized. We note that the comparison made here is based on the assumption that the experimental sample used for measurements is a fully activated, perfect crystal without defects and/or any inaccessibility regions.

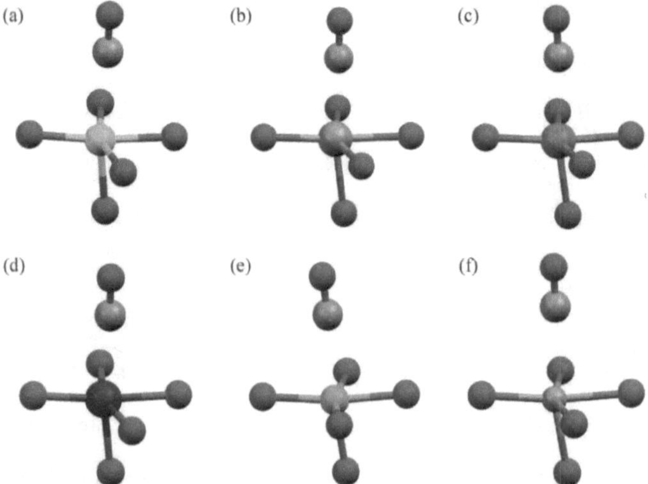

Figure 2.30 Binding geometry of CO adsorbed in (a) Mg-MOF-74, (b) Mn-MOF-74, (c) Fe-MOF-74, (d) Co-MOF-74, (e) Ni-MOF-74, and (f) Zn-MOF-74 computed with the CO ESP-MM developed in this study.

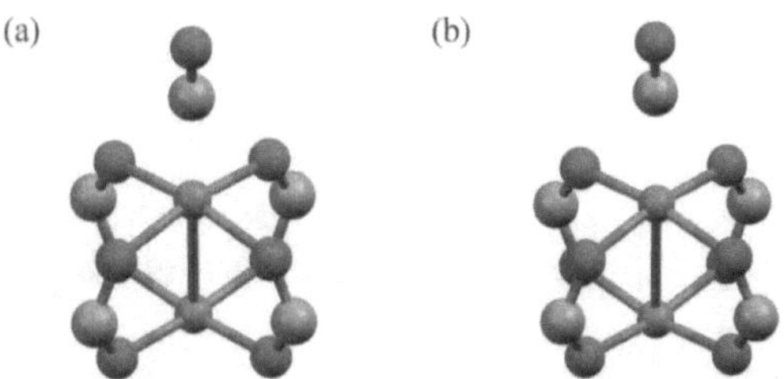

Figure 2.31 Binding geometry of CO adsorbed in Cu-BTC predicted with (a) DFT calculations as reported by Supronowicz et al.[87] and (b) the CO ESP-MM developed in this study.

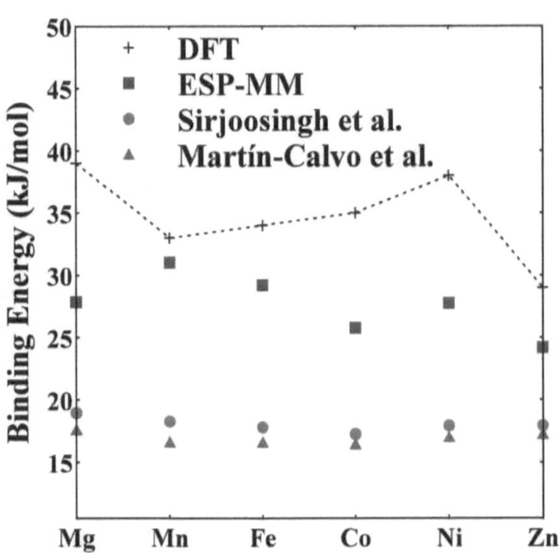

Figure 2.32 Binding energy (kJ/mol) of CO computed with our developed model CO ESP-MM and previously developed CO models[79,82] in M-MOF-74 (M = Mg, Mn, Fe, Co, Ni, and Zn). DFT calculations by Lee et al. is also included for comparison.[27]

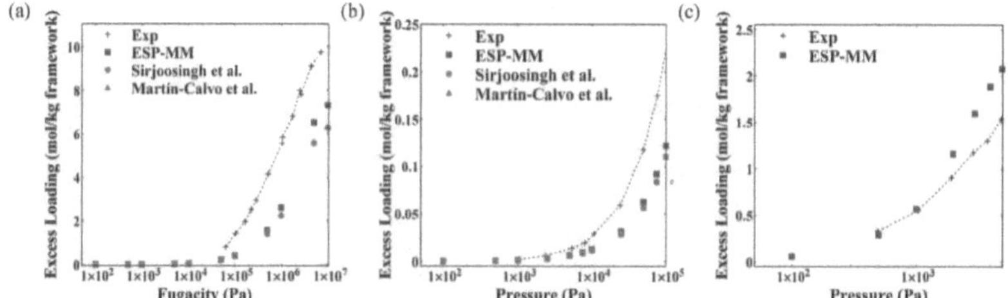

Figure 2.33 Adsorption isotherm of CO in (a) Cu-BTC (295 K), (b) MFI (298 K), and (c) SO$_2$ in MFI (298 K) computed with the ESP-MMs developed in this study, compared with experimental results by Chowdhury et al.[88], Matito-Martos et al.[89], and Deng et al.[90], respectively. Calculations with previously developed models of CO are included.[79,82]

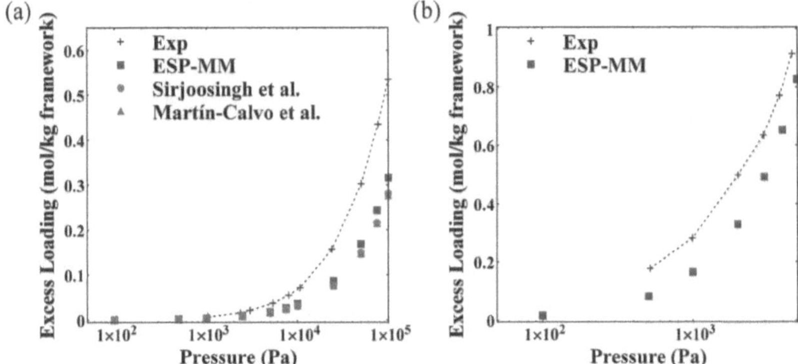

Figure 2.34. Adsorption isotherm of (a) CO (258 K) and (b) SO_2 (323 K) in all-silica MFI computed with the ESP-MMs developed in this study, compared with experimental results by Matito-Martos et al.[89] and Deng et al.[90], respectively. Calculations with previously developed models of CO are included.[79,82]

We also validate ESP-MMs developed in this study for molecules other than CO. For instance, it has been studied that Fe-MOF-74 with N_2O adsorbed is able to activate the C-H bond of ethane and convert it to ethanol. An identification of the binding mechanism of N_2O in Fe-MOF-74 would therefore be advantageous for future applications. Experimentally, it has been reported that there are two binding modes of N_2O adsorbed in the open-metal site of Fe-MOF-74, η^1-O mode and η^1-N mode, with probability of 60 % and 40 %, respectively. With our N_2O ESP-MM, the binding energies at η^1-O mode and

η^1-N mode (Figure 2.35) are predicted to be 32.42 kJ/mol and 31.94 kJ/mol, respectively. Per Boltzmann distribution at room temperature, our calculation predicts 55% and 45% probability, respectively, similar to the experimental results. Furthermore, we compute the binding geometry of other ESP-MMs in Mg-MOF-74 structures, which show consistent results with the previous DFT study by Lee et al. (Figure 2.36).[27] Furthermore, the adsorption isotherms of SO_2 in all-silica MFI at 298 K are calculated and compared with the available experimental data by Deng et al.[90] As can be seen in Figure 2.33 (b), the predictions made by the SO_2 ESP-MM are in good agreement with experiments.

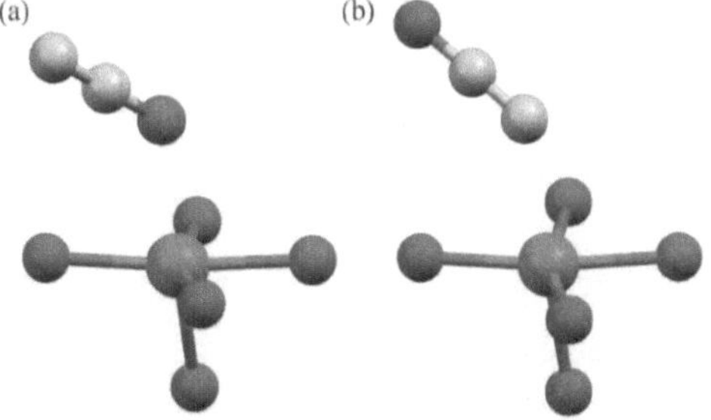

Figure 2.35 Binding geometry of N_2O adsorbed in Fe-MOF-74 with (a) the η^1-O mode and (b) the η^1-N mode predicted with the N_2O ESP-MM developed in this work.

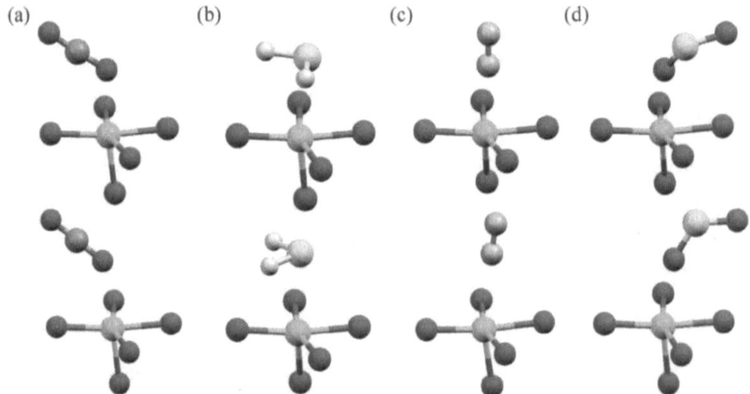

Figure 2.36 Binding geometry of (a) CO_2, (b) H_2S, (c) N_2, and (d) SO_2 adsorbed in Mg-MOF-74 predicted with DFT calculations by Lee et al.[27] (upper) and by the ESP-MMs developed in this work (lower).

To probe whether these ESP-MMs remain a high fidelity in representing ESP when molecules are in close proximity (e.g., relevant configurations in adsorbed phase and in bulk phase), charge determinations using the same methodology as described above are conducted for dimer structures of several selected adsorbates (i.e., H_2S, SO_2, and CO_2; details can be found regarding their dimer structure in Figure 2.37). Table 2.14 shows that the charges assigned to these molecules in their dimer form remain essentially the same (i.e., within an average of ~3% difference). This finding indicates that the assigned charges of our ESP-MMs based on their isolated molecules maintain accurate in representing their ESP when they are in a clustered configuration. We also attempted to conduct the same calculations for H_2O. Surprisingly, the point charges assigned to the water molecule in a dimer structure is found to be drastically different from that in an isolated form. This is an interesting finding, as it implies that the optimal charges assigned to water could in fact be a strong function of its environment and this factor may be important to be considered in the future development of H_2O or other molecules for accurate adsorption predictions.

Table 2.14 Comparison of the point charges assigned to the H_2S, SO_2, and CO_2 molecules in an isolated form (i.e., used to develop ESP-MMs) vs. in a dimer configuration. The same comparison is also conducted for H_2O. The average percentage difference in the assigned charge with respect to the isolated form for each molecule is provided.

	Isolated form (e⁻)	Dimer form (e⁻)	Percentage difference (%)
H_2S	S(-1.152) H(0.268) X(0.308)	S(-1.179) H(0.285) X(0.305)	3.3
SO_2	S(1.077) O(-0.378) X(-0.321)	S(1.102) O(-0.389) X(-0.323)	2.1
CO_2	C(0.6506) O(-0.3253)	C(0.636) O(-0.318)	2.2
H_2O	O(-0.658) H(0.329)	O(-0.802) H(0.401)	21.9

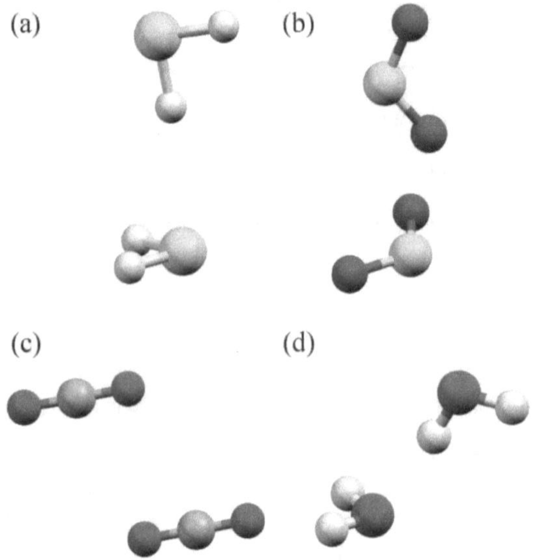

Figure 2.37 Dimer configurations of (a) H_2S, (b) SO_2, (c) CO_2, and (d) H_2O predicted using DFT calculations.

2.2.5 Conclusion

Electrostatic potential optimized molecular models (ESP-MMs) are developed in this study for CO, CO_2, COS, H_2S, N_2, N_2O, and SO_2, of which gas separation or storage are crucially important in energy- or environment-related processes. These molecular models are developed in a systematic and efficient manner. Specifically, the number and location of massless sites and the point charges are firstly optimized to best reproduce the ESP surrounding the molecule, computed by density functional theory (DFT) calculations. Then, the dispersion interaction parameters are fitted to reproduce the experimental vapor-liquid equilibrium. At this step, based on an assumption that the relative magnitude of the Lennard-Jones parameters from generic force fields is sufficiently accurate, two scaling factors are introduced and optimized in order to reduce the overall parameter space. We demonstrated that the developed ESP-MMs from this work are able to accurately reproduce various adsorption properties, as compared to that computed by DFT or measured experimentally. For instance, we show these ESP-MMs, owing to their reliable description in ESP, can accurately predict the binding configuration of molecules in open-metal sites structures, such as M-MOF-74 (i.e., M=Mg, Fe, Co, Ni, and Zn) or Cu-BTC. Furthermore, with ESP-MMs, the predicted adsorption isotherms of CO and SO_2 in Cu-BTC or all-silica MFI are also in decent agreement with experimental measurements. Furthermore, we demonstrated that the accurate ESP representations described with ESP-MMs remain the same when the molecules are in dimer configuration. Overall, our results demonstrate the key role of including ESP during the development of molecular models. We anticipate that these proposed ESP-MMs can be utilized to facilitate the future computational discovery and design of nanoporous materials for various applications.

Chapter 3. Machine Learning-Aided Computational Study of Metal-Organic Frameworks for Sour Gas Sweetening

Reprinted in part with permission from E. H. Cho, X. Deng, C. Zou, L.-C. Lin, *J. Phys. Chem. C* 2020, 124, 50, 27580–27591. Copyright © 2020 American Chemical Society

3.1 Introduction

Metal-organic frameworks (MOFs), a class of nanoporous materials composed of metal nodes and organic linkers, can be promising for H_2S removal from the raw natural gas (i.e., sour gas sweetening).[91] Since the first synthesis more than two decades ago, there have been tens of thousands of MOFs reported to date and orders of magnitude more predicted to be discovered.[14–18] With the vast materials space, identifying optimum MOF candidates is a difficult challenge.

To facilitate the discovery of potential MOFs, classical molecular simulation techniques, such as Monte Carlo or molecular dynamics, have been widely implemented in the past years to screen the vast materials space.[2,10,53,86,92–97] Conducting large-scale studies in a brute-force manner, however, still requires a tremendous amount of computational resources. In light of this, incorporating supervised machine learning algorithms for a significantly more efficient screening study can be a promising direction. Moreover, machine learning methods, such as random forest employed in this work, can also provide quantitative and qualitative analyses on materials' features that result in useful design

106

rules.[28] Machine learning has been employed in making predictions in various fields, ranging from medical diagnosis and treatments[98], natural language processing[99], face recognition[100], spam filtering[101], and other boundless applications. Progresses have also been made in the fields of chemistry and materials science, specifically for adsorption performance of nanoporous materials.[102–105] For instance, Simon *et al.* searched over 670,000 nanoporous structures of a variety of classes for Xenon/Krypton (Xe/Kr) separation.[103] Their study employed the random forest regressor with six geometrical descriptors and one energy-based descriptor for a successful prediction of the Xe/Kr selectivity. Fanourgakis *et al.* also employed the random forest regressor for the prediction of methane adsorption in MOFs, where they involved six standard structural descriptors and four energy-based descriptors.[106]

In this study, we take advantage of machine learning algorithms to study promising structures (i.e., high H_2S/CH_4 selectivity) in the Computational-Ready, Experimentally (CoRE) MOF database[18] for sour gas sweetening and unravel their design principles. We explore aspects that could affect the performance in the predictions, with a specific emphasis on the creation of features that are highly predictive of the H_2S/CH_4 selectivity.[107] Moreover, the features adopted in our study only considers simple-to-compute geometrical and chemical features that can be explicitly interpreted. In previous studies, energy-based descriptors have commonly been included as a feature and are shown to be highly significant for predicting the adsorption or selectivity performance. Despite their predictability, energy-based descriptors generally require more complicated and expensive computations as compared to other geometrical or chemical features and are not transferable to different gas molecules. With the absence of the energy-based descriptor, new features that could take into account the strong adsorption sites of MOFs are needed. In light of this, we herein introduce new geometrical and chemical features that can characterize strong interaction sites that MOFs offer, including the open-metal sites with strong charges and the sites with a high numerical density of framework atoms on the accessible surface. For molecules with strong dipole moments such as H_2S, it is anticipated that the adsorption of H_2S in porous materials is largely influenced by the existence of open-metal sites.[27,31,32,108] Furthermore, the strong interaction sites offered by a high atomic density on the surface are also considered in our study.

In the followings, we firstly explain the methods and details of our study: from data pre-processing, data collection, feature construction, machine learning algorithms, to feature selection. It is then followed by results and discussions, which is divided into three main parts 1) data pre-processing, data collection, and data transformation, 2) feature construction and machine learning modelling, and 3) feature selection.

3.2 Methodology

The procedure of our study follows the steps as schematically shown in Figure 3.1: 1) data pre-processing to remove inaccurate structures, 2) data collection of H_2S/CH_4 selectivity using molecular simulations, 3) feature construction, 4) training/validations of the machine learning model for its predictions of H_2S/CH_4 selectivity of MOFs, and 5) feature selection. Specific details can be seen in the followings.

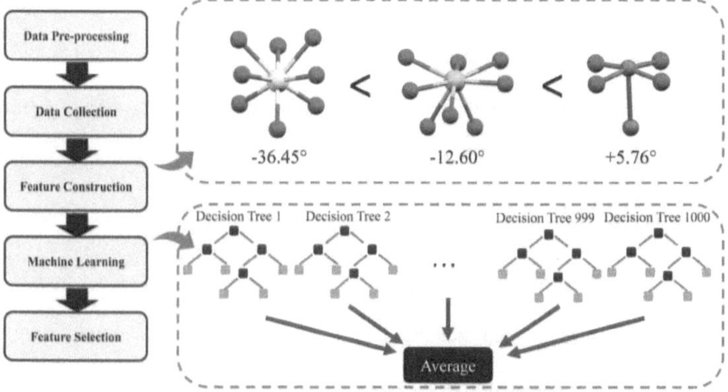

Figure 3.1 Procedure of this study and detailed illustration of the feature construction (i.e., metal angle) and machine learning (i.e., random forest with 1000 decision trees) model. For the feature construction, the metal angle illustrates the geometrical accessibility of the metal node. The most left metal node has the least geometrical accessibility (with metal angle of -36.45°) and the most right metal node has the most geometrical accessibility (with metal angle of +5.76°). The metal nodes illustrated (from left to right) are from HICZEA_clean, GADFAV_clean, and UMUBIP_clean, respectively.

3.2.1 Data Pre-processing

This work studies metal-organic frameworks (MOFs) in the CoRE MOF 2019 database published by Chung *et al.*[18] We consider MOFs in the so-called CoRE MOF 2019-ASR dataset with 360 disordered MOFs excluded. While Chung *et al.* have proceeded curation process for the 11,660 MOFs in CoRE MOF 2019-ASR, as commented by the authors, MOFs with overlapped atoms, partial occupancies, and floating molecules (i.e., ions, solvents, and residuals) may be still observed. Data preprocessing with details shown below is therefore first conducted herein, generally following the procedure reported by Zou *et al.*[109]

Overlapping and Partial occupancy

MOF structures which potentially may possess atoms overlapping or with partial occupancy are detected from the connectivity information. Specifically, we determine that a structure has an overlapping or partial occupancy issue if two adjacent atoms have a distance shorter than a certain threshold of their theoretical bond radii (Table 3.1). A critical threshold value of 0.3 is initially employed to remove the structures with closely grouped atoms. However, MOFs with partial occupancy can have partially occupied atoms separated far away that is over the threshold. Therefore, a stricter (i.e., larger) threshold is determined after the investigation of the distance distribution between the atom pairs of the same element type. For illustration, Figure 3.2 shows the distance distribution between atom pairs of C-C and N-N within the range of bond-distance. Figure 3.2 (a) shows that the highest peak of frequency is located between 1.3 to 1.6 Å for C-C atom pairs, while Figure 3.2 (b) indicates the highest peak of frequency lies in the range of 1.1 to 1.55 Å for N-N atom pairs. According to the Table 3.1, the ideal bond length of C and N pairs should be 1.514 Å and 1.4 Å, respectively, which agrees with the observation in Figure 3.2. The lower bound of these interval is roughly 0.75 of the bond lengths aforementioned, and many of the atom pairs with the distance below this interval are found to have the overlapping and/or partial occupancy issues. We anticipate that such a strict criterion can provide a well-cleaned dataset for the machine learning model, but it is important to note that correct MOFs may also be excluded.

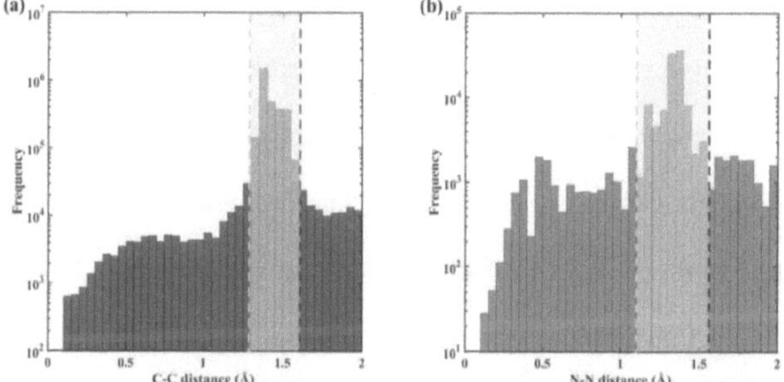

Figure 3.2 Distance distribution of (a) C-C atom pairs and (b) N-N atom pairs. The yellow and black dash lines refer to the lower and higher boundaries of the intervals containing the highest peaks, respectively.

Floating ions or solvents

The detection of floating ions or solvents is also conducted based on the connectivity information of MOFs. Structures with more than one edge-disjoint subgroups (i.e., bonded atoms) are identified and cleaned (i.e., removing floating groups). More details can be found in the work of Zou *et al.*[109]

MOFs with potentially low accuracy of charge assignment

The accuracy of the m-CBAC charges assigned to MOFs may be low if they contain specific atom types that are not included in the training set database of the m-CBAC method.[109] Therefore, such MOFs are also excluded to ensure the best accuracy of the charge assignment.

Repeated and potentially kinetically inaccessible MOFs

Repeated structures with extreme similarities are identified and excluded manually to avoid inappropriate weighted training for machine learning. In addition, since the study of adsorption of H_2S and CH_4 is of the interest, structures with D_f less than 3.6 Å (i.e., the pore limited diameter at which H_2S and CH_4 may be unable to access the framework) are also excluded. Details about the calculations of D_f are shown in Feature Construction Section below.

Overall, amongst 11660 MOFs in CoRE MOF 2019-ASR database, 1327 are removed for potential overlapping and/or partial occupancy concerns, 650 contain elements not included in m-CBAC database, 905 are considered as repeated structures, and 4476 from the rest are calculated as inaccessible.

Table 3.1 Bond radii adopted for identifying the connectivity of an atom. The bond radii from the Universal forcefield (UFF, Rappé *et al.*[49]) are adopted in the m-CBAC approach.[109]

Atom	Ac	Ag	Al	Am	As	At	Au	B	Ba	Be	Bi
Bond Radii ($\AA$)	1.983	1.386	1.244	1.66	1.211	1.545	1.262	0.838	2.277	1.074	1.512
Atom	Bk	Br	C	Ca	Cd	Ce	Cf	Cl	Cl	Cm	Co
Bond Radii ($\AA$)	1.761	1.192	0.757	1.761	1.403	1.841	1.75	1.044	1.044	1.801	1.241
Atom	Cr	Cs	Cu	Dy	Er	Es	Eu	F	Fe	Fm	Fr
Bond Radii ($\AA$)	1.345	2.57	1.302	1.71	1.673	1.724	1.771	0.668	1.335	1.712	2.88
Atom	Ga	Gd	Ge	H	He	Hf	Hg	Ho	I	In	Ir
Bond Radii ($\AA$)	1.26	1.735	1.197	0.354	0.849	1.611	1.34	1.696	1.382	1.459	1.371
Atom	K	La	Li	Lu	Lw	Md	Mg	Mn	Mo	N	Na
Bond Radii ($\AA$)	1.953	1.943	1.336	1.671	1.698	1.689	1.421	1.382	1.47	0.7	1.539
Atom	Nb	Nd	Ne	Ni	No	Np	O	Os	P	Pa	Pb
Bond Radii ($\AA$)	1.473	1.816	0.92	1.164	1.679	1.666	0.634	1.372	1.101	1.711	1.459
Atom	Pd	Pm	Po	Pr	Pt	Pu	Ra	Rb	Re	Rh	Rn
Bond Radii ($\AA$)	1.338	1.801	1.5	1.823	1.364	1.657	2.512	2.26	1.343	1.332	1.42
Atom	Ru	S	Sb	Sc	Se	Si	Sm	Sn	Sr	Ta	Tb

Bond Radii (Å)	1.478	1.064	1.407	1.513	1.19	1.117	1.78	1.398	2.052	1.511	1.732
Atom	Tc	Te	Th	Ti	Tm	U	V	W	Xe	Y	Yb
Bond Radii (Å)	1.322	1.386	1.721	1.412	1.66	1.684	1.402	1.392	1.267	1.698	1.637
Atom	Zn	Zr									
Bond Radii (Å)	1.193	1.564									

3.2.2 Data Collection with Molecular Simulations

Selectivity of H_2S over CH_4 in nanoporous materials are computed with Monte Carlo simulations in a grand canonical ensemble (GCMC) implemented in the RASPA package.[44] The selectivity is defined by

$$S = \frac{x_{H_2S}/y_{H_2S}}{x_{CH_4}/y_{CH_4}} \tag{3.1}$$

where x_i represents the concentration of the adsorbed phase in equilibrium with a bulk phase of mole fraction y_i. The bulk phase composition of sour natural gas varies, and this work considered the composition of the natural gas reserve from Süd-Oldenburg, Germany, whose H_2S:CH_4 ratio is approximately 10:90.[21,110,111] The operating conditions adopted is at 298 K and 5 bar.[112]

The interaction energy is modeled with electrostatic (static point charges) and Lennard-Jones (L-J) potentials. For the MOF frameworks, m-CBAC[109] charges proposed by some of us with the L-J parameters from the universal force field (UFF)[49] are adopted (see Table

3.2). UFF, a generic force field, has been widely adopted in a wide variety of screening studies.[16,113] There have also been some studies that generic force fields are generally sufficiently accurate for screening purposes. For instance, McDaniel et al. assessed the UFF force field we adopted for the frameworks in their predictions of adsorption isotherms for CH_4 and CO_2 in a diverse set of MOFs.[114] They found that the force field is able to predict decently well the ranking of MOFs in their adsorption uptake as compared to accurate quantum mechanically-derived force fields. We also validated the reliability of our simulations by comparing the GCMC-computed isotherms with experimentally reported results for MIL-47(V) (IDIWOH_clean), and both are found in decent agreement as shown in Figure 3.3.[58] For adsorbates, we employ the electrostatic potential optimized molecular model (ESP-MM) for H_2S, considering its accuracy in the representation of electrostatic potential surrounding the molecule (see Table 3.3).[66] The accurate representation of the electrostatic potential has been demonstrated to be especially significant for MOFs that can offer strong electrostatic interaction sites. For CH_4, the TraPPE model is adopted.[115] (see Table 3.4) The Lorentz-Berthelot mixing rule is applied to estimate the L-J parameters between heterogeneous atoms. The L-J potentials are truncated at a cutoff of 12 Å and shifted to zero beyond the cutoff, while electrostatic interactions are computed using Ewald summation with relative error of 10^{-6}. Periodic boundary conditions are implemented and the simulation cell for each MOF has a perpendicular cell width at least twice the cutoff (i.e., 24 Å) along each direction.

Table 3.2 Lennard-Jones parameters employed for framework atoms adopted from universal forcefield (UFF, Rappé *et al.*[49]).

Atom	ε/k_B (K)	σ (Å)
O	30.193	3.118
N	34.722	3.261
C	52.838	3.431
P	153.482	3.695
S	137.882	3.595
Cl	114.734	3.516
H	22.142	2.571
Zn	62.399	2.462
Mn	9.717	2.638
Cu	2.516	3.114
Co	7.045	2.559
Ni	7.548	2.525
Cd	114.734	2.537
Ce	6.542	3.168
Dy	3.523	3.054
Er	3.523	3.074
Gd	4.529	3.001
La	8.555	3.138
Y	36.232	2.980
Ag	18.116	2.805
Al	254.126	4.008
As	155.495	3.769
Au	19.626	2.934
B	90.580	3.638
Ba	183.172	3.299
Be	42.774	2.446
Bi	260.668	3.893
Br	126.308	3.732
Ca	119.766	3.028
Cr	7.548	2.693
F	25.161	2.997
Fe	6.542	2.594

Ga	208.836	3.905
Ge	190.721	3.813
Hf	36.232	2.798
Hg	193.740	2.410
Ho	3.523	3.037
I	170.591	4.009
In	301.429	3.976
Ir	36.735	2.530
K	17.613	3.396
Li	12.580	2.184
Lu	20.632	3.243
Mg	55.857	2.691
Mo	28.180	2.719
Na	15.097	2.658
Nb	29.690	2.820
Nd	5.037	3.185
Np	8.052	3.050
Pb	333.635	3.828
Pd	24.155	2.583
Pr	5.037	3.213
Pt	40.258	2.454
Pu	8.052	3.050
Rb	20.129	3.665
Re	33.742	2.632
Rh	26.671	2.609
Ru	28.180	2.640
Sb	225.946	3.938
Sc	9.561	2.936
Se	146.437	3.746
Si	202.294	3.826
Sm	4.026	3.136
Sn	285.325	3.913
Sr	118.257	3.244
Te	200.281	3.982
Th	13.084	3.025
Ti	8.555	2.829
Tm	3.019	3.006

U	11.071	3.025
V	8.052	2.801
W	33.716	2.734
Yb	114.734	2.989

Table 3.3 Parameters of H_2S ESP-MM employed in this study.

	charge (e⁻)	ε/k_B (K)	σ (Å)	x (Å)	y (Å)	z (Å)
S	-1.152	233.7	3.48	0.00	0.00	0.00
H	0.268	10.3	2.76	-0.97	0.94	0.00
H	0.268	10.3	2.76	0.97	0.94	0.00
X	0.308	0	0	-0.72	-0.50	0.00
X	0.308	0	0	0.72	-0.50	0.00

Table 3.4 Parameters of CH_4 TraPPE model employed in this study.

	charge (e⁻)	ε/k_B (K)	σ (Å)
CH_4	0	148	3.73

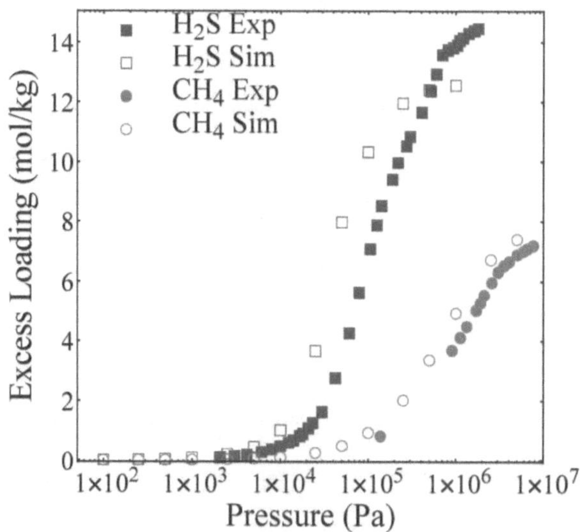

Figure 3.3 Adsorption isotherm of H_2S and CH_4 at 303 K in MIL-47(V) computed with the force field parameters adopted in this study. The computationally-computed isotherm can decently reproduce the experimental result reported by Hamon et al.[58]

A previous study by Anderson *et al.* suggested that the calculations for CO_2/N_2 separation are converged within 3,500 initialization and 3,500 production cycles. Our calculations show that this indeed holds true for a large portion of the structures, particularly those demonstrating low to moderate selectivity (e.g., WODROY_clean as shown in Figure 3.4

(a)). However, for those structures with a large selectivity (e.g., > 250) , calculations may not be converged even with 50,000 initialization and 50,000 production cycles (e.g., HAPVEB_clean as shown in Figure 3.4 (b)). For some of those highly selective structures, the adsorption of CH_4 is very low, leading to difficult samplings. In order to ensure the convergence but at the same time to be practical, we chose to firstly run 5,000 initialization with 5,000 production cycles and subsequently increase more cycles until the results are converged. The production runs are divided into 5 blocks to estimate the mean loadings and the standard error of the mean with ~95 % confidence (i.e., $-2 < z < 2$, where z is the standard score). In order to ensure the convergence of the average, we identified structures whose standard error exceeds 20 % of the estimated mean. For these structures, another 5,000 productions cycles were run restarting from the last cycle and the mean and standard error are then recomputed. This procedure is continued with 10,000, 20,000, 40,000, and 80,000 cycles sequentially (i.e., so that the total production cycles doubles each time) for each structure until convergence is achieved. We note that the maximum number of Monte Carlo cycles is 160,000, considering the available computational resources. However only a negligible fraction of the structures (i.e., ~0.6 %) have standard errors larger than 20% of the estimated mean.

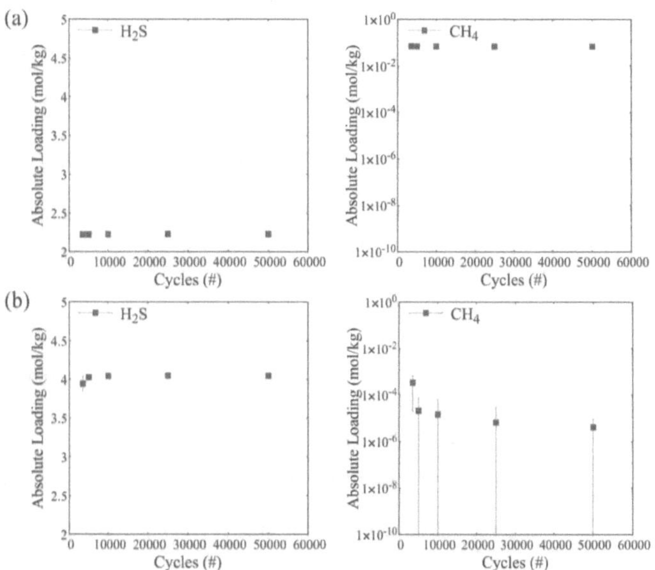

Figure 3.4 Convergence of the computed H$_2$S (left) and CH$_4$ (right) loadings for (a) WODROY_clean and (b) HAPVEB_clean, as a function of the Monte Carlo cycles.

3.2.3 Feature Construction

Machine learning algorithms build mathematical models trained with a subset of sample data (i.e., training set), where each data possess a feature vector and a target value. Predictions can then be made with the model on another subset of sample data, which only have feature vectors and their labels unknown. The features play very significant roles in the predictive power of machine learning models.[116] We incorporate extensively the geometrical and chemical features of MOFs (see Table 3.5 for a list of features). Specifically, several commonly computed geometrical features, including crystal density (ρ), fractional free volume (FFV), accessible gravimetric and volumetric surface area ($gASA$ and $vASA$), largest cavity diameter (D_i), and pore limiting diameter (D_f) are computed. From the computed D_i and D_f, their ratio D_i/D_f are also included in the feature.[103] Furthermore, the most negative charge (MNC) of a given structure, a chemical feature adopted in a previous study, is considered.[117] On top of them, we develop new features to improve the predictability of machine learning model. Specifically, for each structure of the dataset, the accessibility of the open-metal sites as well as their strong electrostatic interaction are quantified. These include the metal openness angle, the charge of the metal, and the dipole moment of the metal node. Another geometrical feature, the surface atom density (ρ_{AS}), is also designed in this study. For a given surface area, gas adsorbates could effectively interact strongly with the adsorbent if the atoms are more compact on the surface of the pores. All of these features are quantified using an in-house script. In the followings, specific details regarding the features are shown.

Table 3.5 Features considered in this study for training machine learning algorithm. Note that the order does not indicate their relative importance.

Feature	Description
metal angle ($\textbf{MA}$) [°]	maximum average projection angle of metal in a given structure
metal charge ($\textbf{MC}$) [e]	charge of the metal node with the maximum average projection angle of metal in a given structure
metal dipole moment ($\textbf{MDM}$) [eÅ]	dipole moment of the metal node with the maximum average projection angle of metal in a given structure
surface atom density ($\boldsymbol{\rho_{AS}}$) [m^{-2}]	numerical density of framework atoms on the accessible surface
most negative charge ($\textbf{MNC}$) [e]	most negative charge of a given structure
crystal density ($\boldsymbol{\rho}$) [g/cm^3]	mass of the framework atoms per volume
fractional free volume ($\textbf{FFV}$) [-]	ratio of space that can be accessed by the Helium (He) gas and the volume of the unit cell
accessible gravimetric and volumetric surface area ($\textbf{gASA}$ and $\textbf{vASA}$) [m^2/g and m^2/cm^3]	area of framework atom surfaces that can be accessed by the Argon (Ar) gas
largest cavity diameter ($\boldsymbol{D_i}$) [Å]	diameter of the largest sphere that can be positioned in the unit cell without overlapping with the structural atoms
pore limiting diameter ($\boldsymbol{D_f}$) [Å]	diameter of the narrowest part of the diffusion pathways
D_i/D_f ($\boldsymbol{D_i/D_f}$) [-]	ratio of D_i and D_f

Metal Angle (MA), Metal Charge (MC), and Metal Dipole Moment (MDM)

It has been demonstrated in previous studies that open-metal sites offer strong interaction sites for gas molecules with strong dipole or quadrupole moments, such as CO_2, H_2S, SO_2, and NH_3.[27,31,32,108] These open-metal sites are created in the process of the synthesis and activation of MOFs; during the synthesis, the metal centers coordinate to both solvent molecules and linkers; following an activation process where the solvents are removed, coordinatively unsaturated open-metal sites are formed.[118] Recently, the authors of the CoRE MOF 2019 has provided an algorithm for binary classification that can distinguish open-metal sites.[18] While their algorithm can detect OMS metals, we herein propose an approach to further quantify the geometrical openness or accessibility of the metal sites. Firstly, the first coordination sphere of a given metal is detected, which serves as vertices of an irregular polyhedron (see Figure 3.5 (a) for Zn-MOF-74). For each plane face of the polyhedron created (i.e., 6 plane faces for the case of Zn-MOF-74 as shown in Figure 3.5 (b)), the extent of how much the metals are embedded inside the polyhedron or how much the metals are projected out of the polyhedron is computed. In these calculations, the higher values indicate greater openness or accessibility of the metal nodes. Among the values computed for each plane face, the plane face with the highest values is selected (+4.09° for the case of Zn-MOF-74, see Figure 3.5 (b)).

The calculations of the feature "metal angle" for a single metal node for the Zn-MOF-74 as an example are as follows:

1. First coordination sphere of a given metal are identified (i.e., 5 oxygen atoms). Two atoms are considered bonded if the interatomic distance is below 1.25×(sum of the atomic radii). The atomic radii definitions can be found above in Table 3.1.

2. An irregular polyhedron is created with the first coordination sphere atoms (i.e., 6-plane-face irregular polyhedron, see Figure 3.5).

3. For the three atoms that constitutes a given plane face, lines that connect the metal and each of the three atoms are generated to compute the angle between the line and its orthogonal projection onto the plane.

4. a. If the metal is located inside the polyhedron, the angle (which we denote as ∠[atom1, atom2, atom3]) is calculated as the negative value of the average of the three acute angle computed in step 3 (i.e., ∠[O1, O3, O4], ∠[O1, O2, O3], ∠[O2, O3, O5], and ∠[O3, O4, O5], see Figure 3.5).

    b. If the metal is located outside of the polyhedron, the metal openness is calculated as the positive value of the average of the three acute angle computed in step 3 (i.e., ∠[O1, O2, O4] and ∠[O2, O4, O5], see Figure 3.5).

5. Step 3 and 4 are repeated for all the plane faces of the irregular polyhedron and the largest metal angle is selected as the one that represents the considered metal (i.e., ∠[O2, O4, O5]: +4.09°, see Figure 3.5).

We note that in step 2, if there are less than two atoms in the first coordination sphere, metal angle is assigned as +65°, and if there are three atoms, only one plane, rather than a polyhedron, is generated and the average projection angle regarding the plane is computed. Since the upper limit of the metal angle computed for metal nodes with more than three atoms in the first coordination sphere is computed to be +65°, setting this value as the metal angle for the cases where the metal atoms is bonded to only two atoms enables the feature "metal angle" to be continuous and non-discrete.

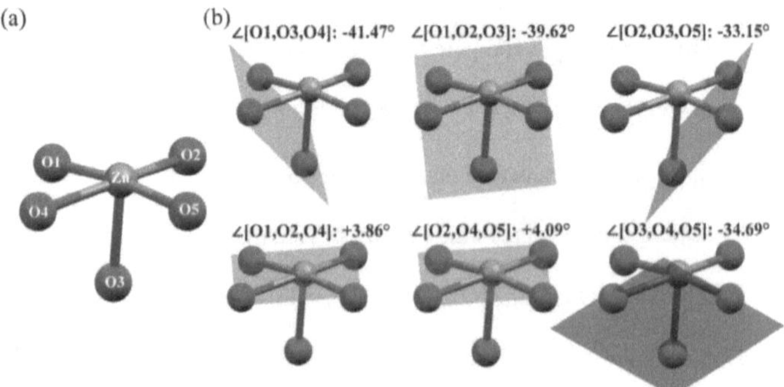

Figure 3.5 A schematic illustration of the (a) first coordination atoms of Zn of Zn-MOF-74 (FIJDOS_clean) and (b) average projection angle for each plane face of the polyhedron created by the first coordination atoms. In (b), the middle of the bottom row, which has the largest angle, is selected as the representative angle for Zn-MOF-74.

We note that, for a given material, the metal angle is computed for all possible metal nodes. Among the metal nodes which can either be homogeneous (i.e., same type) or heterogeneous (i.e., multiple types) throughout a given structure, the metal node that has the highest metal angle is selected as the feature "metal angle" (MA) for that particular structure. The metal charge (MC) and metal dipole moment (MDM) of the same metal node that represents the electrostatic field are also computed and included as features.

MNC

The most negative charge assigned to atoms in a given structure.

Fractional Free Volume (*FFV*)

In this study, the fractional free volume (*FFV*) refers to the ratio of space that can be accessed by the Helium (He) gas ($V_{accessible\ to\ He}$) and the volume (V) of the unit cell. In these calculations, the unit cell is meshed into evenly-distributed grid points with a grid distance of less than 0.2 Å along the cell directions. The shortest distance between each grid point and the nearest framework atom excluding the van der Waals radius (r_{vdW}, see Table 3.6) was then calculated. This distance represents the radius of the largest sphere (r_{LS}) that can be positioned at the said grid point without overlapping with the framework atoms. This process is illustrated in Figure 3.6. The FFV is calculated by ratio of grid points that can accommodate a He atom ($r_{vdW,He}$=1.40 Å) as shown in Equation 3.2.

$$FFV = \frac{Number\ of\ grid\ points\ that\ can\ accommodate\ a\ He\ atom\ (i.e.,\ r_{LS} > r_{vdW,He})}{Number\ of\ grid\ points\ in\ the\ unit\ cell} \tag{3.2}$$

Table 3.6 Van der Waals radii (r_{vdW}s) of framework atoms used in the geometrical feature calculations adopted from Rappé *et al.*[49]

element	Ag	Al	As	Au	B	Ba	Be	Bi	Br	C	Ca	Cd	Ce
r_{vdW} (Å)	1.40	2.00	1.88	1.47	1.82	1.65	1.22	1.95	1.87	1.72	1.51	1.27	1.58
element	Cl	Co	Cr	Cs	Cu	Dy	Er	Eu	F	Fe	Ga	Gd	Ge
r_{vdW} (Å)	1.76	1.28	1.35	2.01	1.56	1.53	1.54	1.56	1.50	1.30	1.95	1.50	1.91
element	H	Hf	Hg	Ho	I	In	Ir	K	La	Li	Lu	Mg	Mn
r_{vdW} (Å)	1.29	1.40	1.21	1.52	2.00	1.99	1.27	1.70	1.57	1.09	1.62	1.35	1.32
element	Mo	N	Na	Nb	Nd	Ni	Np	O	P	Pb	Pd	Pr	Pt
r_{vdW} (Å)	1.36	1.63	1.33	1.41	1.59	1.26	1.53	1.56	1.85	1.91	1.29	1.61	1.23
element	Pu	Rb	Re	Rh	Ru	S	Sb	Sc	Se	Si	Sm	Sn	Sr
r_{vdW} (Å)	1.53	1.83	1.32	1.30	1.32	1.80	1.97	1.47	1.87	1.91	1.57	1.96	1.62
element	Tb	Te	Th	Ti	Tm	U	V	W	Y	Yb	Zn	Zr	
r_{vdW} (Å)	1.54	1.99	1.51	1.41	1.50	1.51	1.40	1.37	1.49	1.49	1.23	1.39	

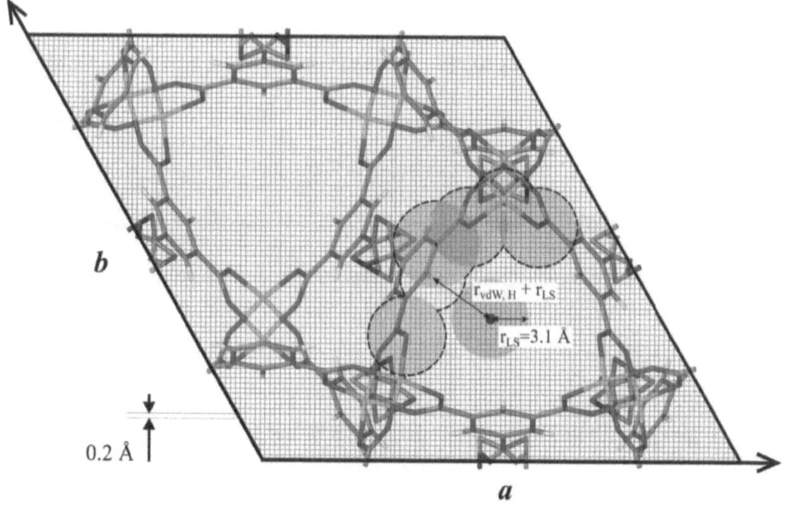

Figure 3.6 Two-dimension (viewed along the c direction) illustration of grid points of the WUGQAS structure. For the grid point highlighted (blue dot), the nearest framework atom was a H atom as illustrated by the contour of nearby structural atoms (black dashed line). The radius of the largest sphere (r_{LS}) that can be positioned at the highlighted grid point is calculated by deducting the distance between the H and the grid point (4.4 Å) by $r_{vdW,H}$ (1.3 Å) as 3.1 Å. Color code: C, grey; H, white; O, red; and Zn, cyan.

Gravimetric and Volumetric Accessible Surface Area (gASA and gASA)

In this study, the accessible surface area (ASA) is defined as the area of framework atom surfaces that can be accessed by insertion of the Argon (Ar) gas ($r_{vdW,Ar}$=1.88 Å). For each framework atom, 5,000 trial points are randomly collected from its concentric sphere with a radius of r_{vdW} of the framework atom and Ar combined. For each of the trial points, the r_{LS} is calculated to determine whether it can accommodate an Ar atom. Following this scheme, the ASA contribution of the framework atom i is calculated as shown in Equation 3.3.

$$ASA_i = \frac{Number\ of\ trial\ points\ that\ can\ accommodate\ an\ Ar\ atom}{5{,}000} \cdot 4\pi(r_{vdw,i} + r_{Ar})^2 \qquad (3.3)$$

$$ASA = \sum ASA_i \qquad (3.4)$$

To ASA is presented in the commonly-used units of m²/g and m²/cm³ by dividing the ASA obtained in Eqn. 3 with m, which is the total mass of all framework atoms, and V, respectively.

Largest cavity diameter (D_i)

The largest cavity diameter (D_i) refers to the diameter of the largest sphere that can be positioned in the unit cell without overlapping with the structural atoms. The D_i is calculated as the twice the largest r_{LS} of the unit cell as shown in Equation 3.5.

$$D_i = 2\,r_{LS} \qquad\qquad (3.5)$$

Pore limiting diameter (D_f)

To quantify the pore limiting diameter (D_f), starting from 0 Å, probe with increasing diameters at an increment of 0.2 Å are determined whether they can travel across the unit cell starting from any of the grid points (i.e., reach the periodic image of the grid point from which the probe started) by jumping between adjacent grid points by employing the flood fill algorithm. The collection of grid points that can be visited by the probe from the starting grid point consequently constitutes the diffusion channel. The diameter of the largest probe that can travel across the unit cell (i.e., generate connected diffusion channels) is therefore the D_f of the structure. The process of locating D_f is illustrated in Figure 3.7.

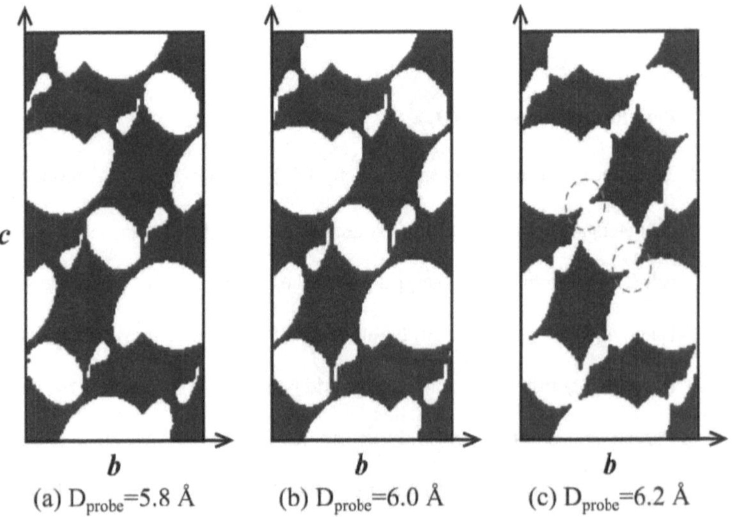

Figure 3.7 Illustration of the process of determining the diameter of the narrowest part of the diffusion channels (D_f) using the WUGQAS structure as an example. The tiny blue balls represent grid points that can accommodate probes at the specified grid diameters (D_{probe}). As shown, in panels (a) and (b) (D_{probe} is 5.8 and 6.0 Å, respectively) where the grid points are interconnected and therefore connected diffusion channels can be established, the probe can travel across the unit cell. However, as exemplified by the highlighted parts (enclosed in red dotted circles) of panel (c), discontinuity occurs, and the diffusion channel is broken. The D_f is, therefore, determined as 6.0 Å.

For the geometrical features, the framework atoms are modeled as hard spheres with sizes defined by the van der Waals radii (r_{vdW}, see Table 3.6) defined in UFF.[49] The unit cell is then meshed into fine grids with a grid distance of smaller than 0.2 Å along cell directions (i.e., *a*, *b*, and *c*) to sample the aforementioned features (see Figure 3.6).

Surface Atom Density (ρ_{AS})

The surface atom density (ρ_{AS}) refers to how densely-packed the framework atoms are on the accessible surface. The so-called surface atom density is calculated as the number of framework atoms with non-zero surface area contribution (i.e., $N_{ASA\neq0}$) divided by the total accessible surface area (*ASA* [m^{-2}], see above), as shown in Equation 3.6:

$$\rho_{AS} = \frac{N_{ASA\neq0}}{ASA} = \frac{N_{ASA\neq0}}{(vASA \times volume)} \tag{3.6}$$

where $N_{ASA\neq0}$ denotes the number of framework atoms with non-zero contribution to the *ASA* and *vASA* represents the volumetric accessible surface area (m^2/cm^3). A high density of surface atoms is anticipated to provide stronger interactions with adsorbates.

3.2.4 Learning Algorithms

Among the nonlinear supervised machine learning models, we choose to employ the random forest regressor, an *ensemble* of decision trees, which have been widely adopted in many applications, ranging from remote sensing[119,120], gene selection,[121] land-cover monitoring,[122] disease risk predictions,[123] to computer vision.[124] A schematic illustration of a single decision tree regressor is shown in Figure 3.8.

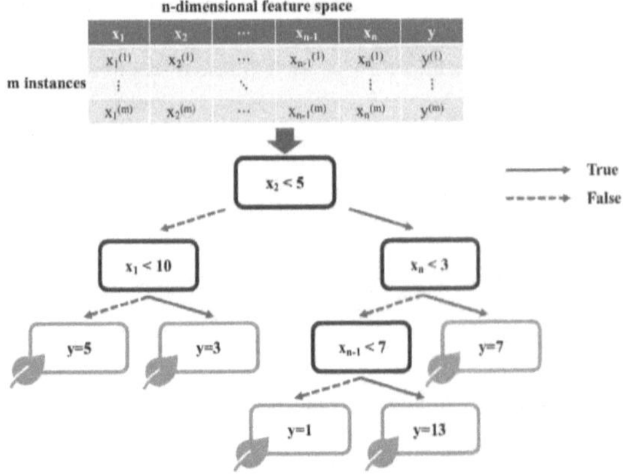

Figure 3.8 A schematic illustration of a single decision tree. Leaf nodes are marked and colored as orange.

A decision tree consists of a root node (the most upper node) where training data comes in of the form $(x^{(i)}, y^{(i)})$ for $i = 1, 2, \cdots, m$, where $x^{(i)} \in \mathbb{R}^n$ and $y^{(i)} \in \mathbb{R}$. Then, the root node is split into two child nodes, which satisfies $Q_{left}(j, t) = (x, y)|x_j < t$ and $Q_{right}(j, t) = (x, y)|x_j \geq t$, respectively. The trees are grown by subsequently splitting each of the child nodes until no splits are possible. The last nodes are denoted as leaf nodes and each leaf node is labeled by an average of the target values inside the given leaf node. At each decision node k, the choice of (j_k, t_k) is determined as the one that most minimizes the weighted average of impurity, which is defined by:

$$G(Q, j_k, t_k) = \frac{n_{left}}{N_k} H\left(Q_{left}(j_k, t_k)\right) + \frac{n_{right}}{N_k} H\left(Q_{right}(j_k, t_k)\right) \qquad (3.7)$$

where the $N_k = n_{left} + n_{right}$ is the total number of data at node k and $H()$ is the impurity function. For regression, the commonly used impurity function for measuring the quality of a split is mean squared error (MSE), which is defined as:

$$MSE = \frac{1}{m} \sum_{i=1}^{m} (y_i - \mu)^2 \qquad (3.8)$$

$$\mu = \frac{1}{m} \sum_{i=1}^{m} y_i \qquad (3.9)$$

where y_i for i=1, ... , m indicates the target variable of the ith instance and μ is an average of the target variable for m instances.

In our work, 1000 trees are employed for training random forest learners and no restrictions are imposed so that the trees are grown to their full extent. The scikit-learn package of Python is employed herein for the training of the random forest regressor and the predictions.[125]

3.2.5 Feature Selection

It is critical to identify important features that show high predictability of the H_2S/CH_4 selectivity. As decision trees or random forests suffer relatively less from overfitting caused by unimportant features as compared to other models (e.g., deep neural nets)[107,116], we herein explore an exhaustive number of features and then evaluated the significance of each feature using various feature selection methods. The impurity-based feature importance method and wrapper feature selection methods, specifically the forward selection, the backward elimination, and the exhaustive search, are adopted.[126] In the impurity-based feature importance method, the feature, which is selected in a decision node, that can reduce the impurity of the child nodes the most are considered to be the most important one. Followings are details about the wrapper feature selection methods.

Forward Selection: The forward selection method proceeds by firstly selecting a single feature for building a random forest regressor. This is done for every feature in Table 3.5 (i.e., 12 features) and the feature that yields model with the best performance is selected. In the next step, from the remaining features (i.e., 11 features, excluding the one selected in the previous step), another feature is selected for building a random forest regressor, together with the feature selected in the previous step (i.e., build random forest with 2 features). Among 11 features, the one that produces the best performing model is selected. This procedure is continued until all the features are exhausted. The order which the features are selected can be considered as an indicator of the importance of the features.

Backward Elimination: In contrast to the forward selection approach, backward elimination method progresses by firstly building a random forest regressor with the full set of features. In the first step, each of the features is removed to build new models. Among the features, the one whose exclusion yields the best performance model is considered to be the least important feature and is eliminated in further steps. With the remaining features (i.e., 11 features), each of the remnant features are removed to build new models and the feature whose removal results in the highest performing model is eliminated in the later steps. This process is continued until only one feature remains.

Exhaustive Search: One can also conduct a non-greedy, thrifty process of feature selection that does not depend on what features are selected in other steps, through an exhaustive search of all the possible combinations of the features. This process is conducted for varying number of features. For instance, if only one feature is allowed, all the possible 12 cases (i.e., $_{12}C_1 = 12$) are considered for the modelling and the one with the best performance is chosen. If n features are allowed, all the $_{12}C_n$ cases are considered and the one with the lowest root mean square error (RMSE) is chosen.

3.3 Results and Discussion

3.3.1 Data Pre-Processing, Data Collection, and Data Transformation

Among structures after data pre-processing, approximately 1600 structures are randomly selected without replacement to compute the H_2S/CH_4 selectivity and build/validate a random forest regressor. The computed selectivity of these structures show heavily positively-skewed distribution (Figure 3.9 (a)). Although normalization/standardization of the features do not affect performance of a random forest regressor as the paths from the root node to the leaf node follow classification rules, the non-normal distribution of the target value can detriment the fitness of models and thus lead to unreliable modelling,[127] which will also be discussed later. For this reason, a box-cox transformation, a family of power transformation that is often used to transform variables with skewed distribution to gaussian-like distribution, is applied to the target variable selectivity. The selectivity is box-cox transformed by the following equation:

$$y(\lambda) = \begin{cases} \dfrac{y^{\lambda} - 1}{\lambda}, & if\ \lambda \neq 0; \\ \log y, & if\ \lambda = 0. \end{cases} \qquad (3.10)$$

where λ is optimized by the maximum likelihood estimation. The λ for our datasets are optimized as -0.23. The distribution of the transformed selectivity is shown in Figure 3.9 (b). For the transformed datasets, statistical tests for the skewedness and kurtosis are conducted. The statistical hypothesis test for skewedness concludes a p-value of 0.24, which with significance level of 0.05, fails to reject that the skewedness is same as that of the corresponding normal distribution. The p-value from the kurtosis test is 1.7×10^{-6}, which with a significance level 0.05, rejects that the kurtosis is same as that of the corresponding normal distribution. Namely, the distribution is symmetrical and gaussian-like, despite its long tails at the lower and upper bounds. However, our manual inspection did not identify them as outliers, which also is known as the *"black swans effect"* as proposed by Taleb.[128] Thus, we adopt all the box-cox transformed data for training/validating machine learning model.

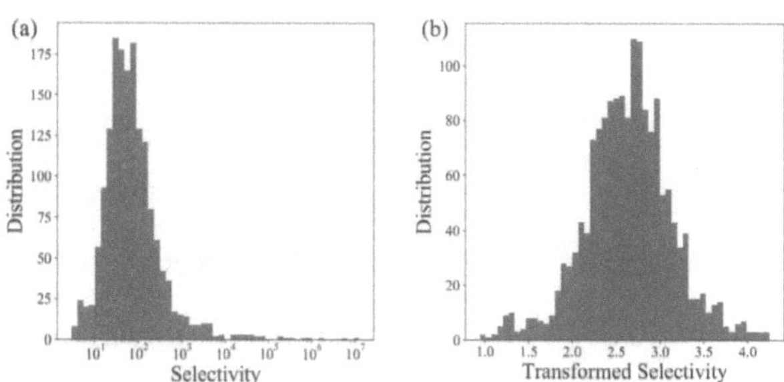

Figure 3.9 Histogram of the (a) selectivity and the (b) box-cox transformed selectivity. Note that (a) is plotted in logarithm scale for better visualizations.

3.3.2 Feature Construction and Machine Learning Modelling

Feature Construction

As mentioned above, we incorporate newly designed features as well as standard geometrical and chemical features that have been commonly employed in previous studies for building machine learning model. Figure 3.10 shows the correlations of the selectivity with each of the features. It shows that structures with high selectivity indeed either have open-metal sites with strong charges (Figure 3.10 (a) and (b), structures that have $MA > 0°$ and $MC > 1.0$ e, respectively), an optimal range of dipole moment (Figure 3.10 (c), MDM of 2–5 eÅ). Highly selective structures tend to also have high ρ_{AS} (Figure 3.10 (d)). The selectivity is also shown to be highly correlated with standard geometrical features. Large pore sizes generally tend to result in low selectivity. This can be seen from that highly selective structures exhibit lower fractional free volume (FFV), lower largest cavity diameter (D_i), and pore limiting diameter (D_f). For these structures, H_2S can interact strongly with framework atoms via interactions and occupies the pore centers, thus precluding CH_4 from adsorption and allowing a high selectivity. The correlation matrix of the features is shown in Figure 3.11. Some of these features show high correlations to each other. Specifically, many of the geometrical features commonly adopted previously are highly correlated to each other. For instance, FFV has high correlations with $gASA$, D_i, and D_f, with their correlation coefficient exceeding 0.7 (see Figure 3.11). Interestingly, D_i/D_f, a feature that could inform the channel shape for MOFs with a relatively simple topology, is not strongly correlated with any of the features considered, with only a moderate correlation with D_i. MNC also shows very little correlations with other features, including the MC. ρ_{AS}, a new feature, is highly correlated with $vASA$ (correlation coefficient of -0.76) and moderately correlated with ρ, FFV, $gASA$, and D_i (correlation coefficient of 0.41, -0.66, -0.63, and -0.46, respectively). On the other hand, other new features that accounts for the open-metal sites with strong electrostatic field (i.e., MA, MC, and MDM) show weak correlations with other features, with an exception between MC and MDM (correlation coefficient of 0.5). This indicates that the newly designed features introduce new information, specifically the local geometrical and chemical information of the metal nodes that cannot be described by the standard geometrical and chemical features.

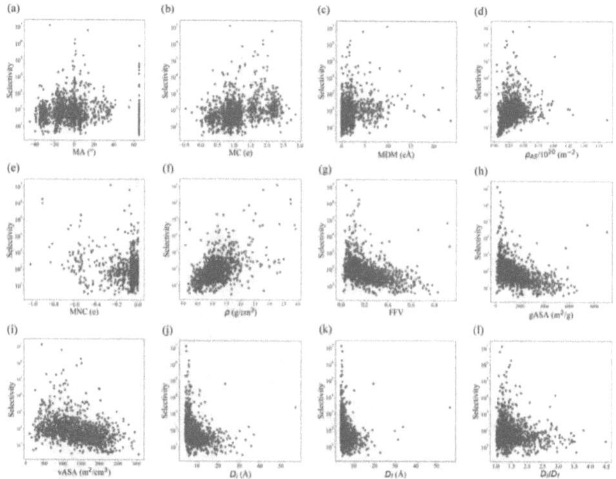

Figure 3.10 Correlation between selectivity and each feature considered in this study. The order of the features are same as that in Table 3.5 and does not indicate their relative importance.

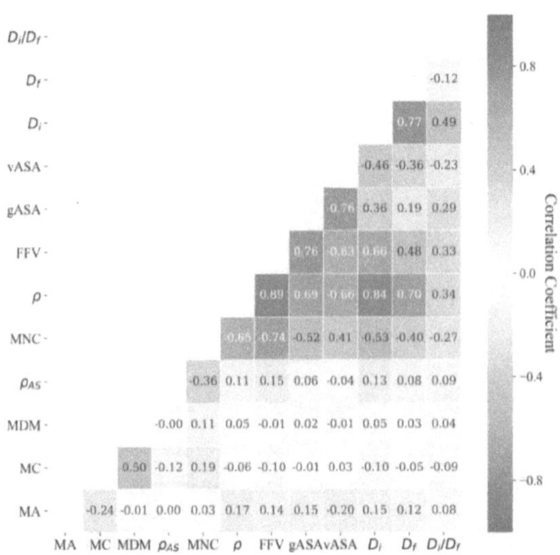

Figure 3.11 Correlation matrix of the features considered in this study. Each grid is colored and annotated by the correlation coefficients of the explanatory variables. The features are denoted with their symbols defined in Table 3.5.

Machine Learning Training/Validation

A random forest regressor is trained with all features discussed above. Despite high correlations between some of the features, random forests are especially immune to overfitting due to irrelevant or highly correlated features. Among the ~1,600 structures with known labels, 80% are used for the training and the remaining are used for validation. The number of the training sets is sufficient for reliable predictions as can be seen from the learning curve shown in Figure 3.12. The forest is constituted of 1000 trees, which is also shown to be sufficient for reliable prediction (see Figure 3.13). The performance of the trained model can be seen in Figure 3.14 (e) and (f), with an root mean square error (RMSE) of 0.1189 and 0.3206 for the training set and the test set, respectively. This is a noticeable improvement as compared to when only involving conventional geometrical and chemical features. Specifically, when using only the standard geometrical and chemical features, the RMSE on the same training set and the validation set are 0.1375 and 0.3821, respectively (See Figure 3.14 (a) and (b)). As shown in Figure 3.12, this RMSE of 0.3821 with the standard features (trained with ~1300 datasets) is similar to that trained with the whole set of features including the newly designed ones but only with ~130 datasets (i.e., 10%). In other words, the introduction of the descriptors that features the open-metal site and the atom density on the pore surface of MOFs has the effect of using 90% fewer datasets for the training, which is significant considering the computational resources required to obtain each data. We expect that the importance of the newly designed features would be more pronounced for systems where the composition of H_2S is higher than 10 % (i.e., composition of the current study), as the open-metal sites are offering strong interaction sites to gas molecules that have strong dipole moments. One should be noted, however, that the standard features are still useful, but rather lack some important information regarding local geometrical and chemical information of the metal nodes that are relatively more open to the gas adsorbates. Indeed, these standard features contain essential information regarding the global geometrical features of a given structure. Therefore, with the absence of these standard features, the performance in the predictions is not as good as that of using full set of features. Specifically, the performance of machine learning trained only with the new features (i.e., without the standard features) can be seen from Figure 3.14 (c) and (d), where the RMSE of the training and validation sets are 0.1327 and 0.3382, respectively. The performance is better than when only using standard features but worser than when using the full set of features.

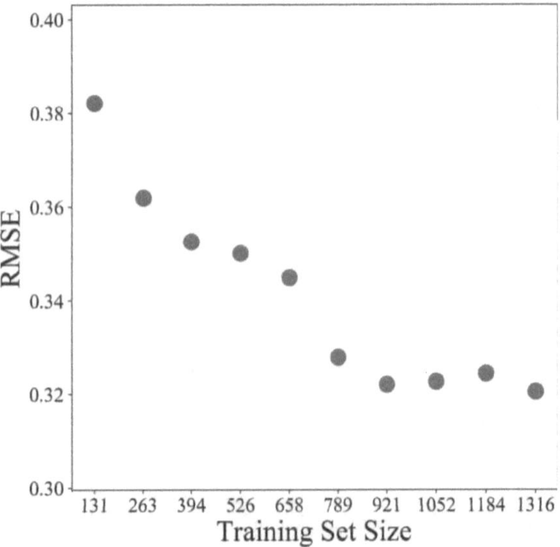

Figure 3.12 Performance of random forest regressor on the validation set with different training set sizes.

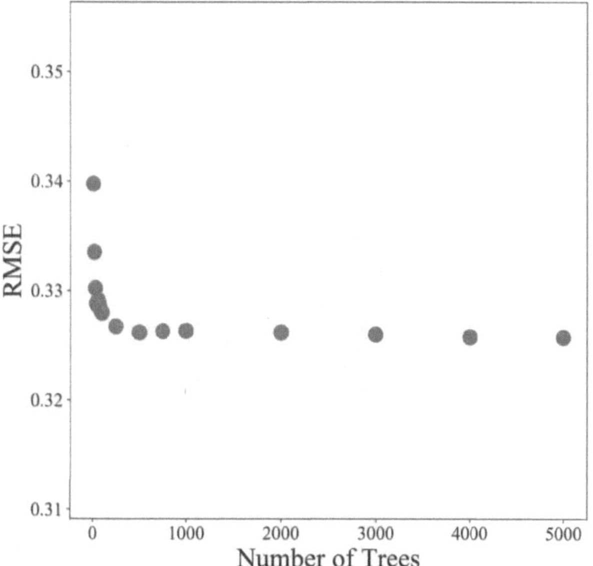

Figure 3.13 Performance of machine learning with respect to varying number of decision tree estimators. 10-fold cross validation was employed for each number of trees.

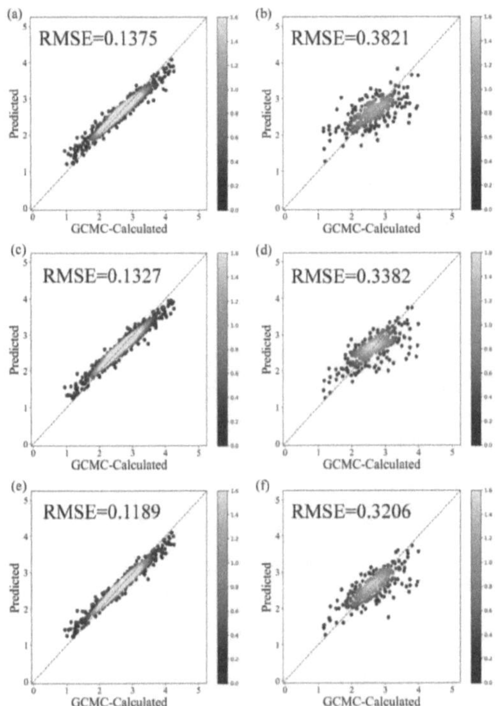

Figure 3.14 Random forest trained with standard geometrical and chemical features (top row, a-b), newly designed features (middle row, c-d), and the total set of features (bottom row, e-f), with predictions on training set (left column) and validation set (right column). The x-axis represents the GCMC-calculated selectivity while the y-axis shows the random forest-predicted one. Warmer color indicates higher density of the data. Root mean square error (RMSE) is shown for each case.

At this point, it is important to highlight the necessity of the transformation of the label for training and predictions. With the same training and validation data sets without the box-cox transformation (i.e., selectivity) for training/validating random forest regressor, the fitting and predictions are highly inaccurate, as shown in 3.15. Especially, the errors for the training sets are even higher than that of the validation sets, indicating a poor fitting with the highly skewed datasets. By employing the box-cox transformation, the current model demonstrates a good fitting of the training set and a good prediction of the validation set as shown above.

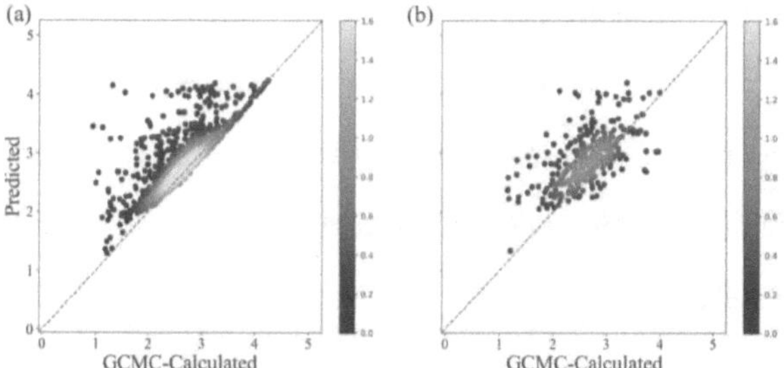

Figure 3.15 Random forest predictions on (a) training set and (b) validation set trained with original datasets (i.e., before box-cox transformation). x-axis represents the GCMC-calculated selectivity while the y-axis shows the random forest-predicted one. Warmer color indicates higher density of the data. The datasets are box-cox transformed after training/validation for better comparison and visualization.

3.3.3 Feature Selection

Impurity-based Feature Importance

Random forest can be advantageous for feature engineering as it can estimate impurity-based relative importance of the features. For the features that are considered in the decision node, those that can reduce the impurity of the child nodes more are considered to be more significant. From our model, the most important feature estimated is FFV, followed by the features that capture the strong bonding sites of the open-metal sites (i.e., MA, MC, and MDM) and ρ_{AS} (see Figure 3.16). Interestingly, the least important features turn out to be those commonly used geometrical features, such as D_i, D_f, $vASA$, and D_i/D_f. This result indicates the important roles of the newly designed features. However, we note that MA, MC, and MDM contain only the local geometrical and chemical information of the metal nodes and lack information about the overall structural properties of the pores of the structure. For this reason, some other features such as FFV or ρ_{AS} are still needed for the modelling.

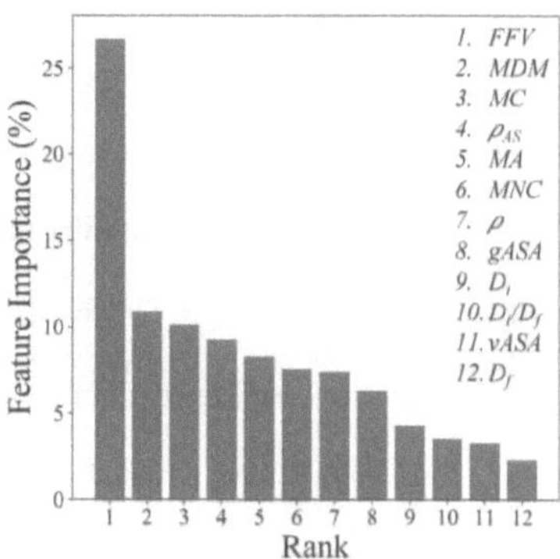

Figure 3.16 Relative importance of the features (impurity-based) considered in this study, ordered by their importance.

Wrapper Feature Selection Methods

We also implement wrapper feature selection methods, including the backward elimination and forward selection. Unlike the impurity-based feature importance which is based on how the decision trees grow, these methods are based on how a given feature contributes to the performance on the validation set. 10-fold cross validation is used to assess the performance. Specific details about the procedures can be seen in the Methodology section above. The results of the backward elimination and forward selection are shown in Figure 3.17. Our newly developed features, such as MDM, MA, MC, and ρ_{AS}, are consistently ranked high. MNC is also determined as one of the most important features. Interestingly, FFV is considered to be the least important in the backward elimination and at the same time the most important in the forward selection. Similarly, density is identified to be the most important in the backward elimination but one of the least important in the forward selection. This could be attributed to the greedy algorithm of these methods (i.e., once a feature is selected for a forward selection, it should still be included in the proceeding steps) and at the same time high correlations between features. For instance, while FFV is shown to be significant from the impurity-based feature importance, it is highly correlated with other features and therefore largely compensated by other geometrical features. As mentioned earlier, the correlation coefficient between FFV and $gASA$ or D_i can be as high as 0.89 and 0.84, respectively (Figure 3.11). Nonetheless, certain standard geometrical features are identified as less important than other features in both the backward elimination and the forward selection, such as D_i/D_f, D_f, $gASA$, D_i, and $vASA$, as their removal or addition do not noticeably affect the performance of machine learning. These features are also shown to be relatively less important from the impurity-based feature importance method.

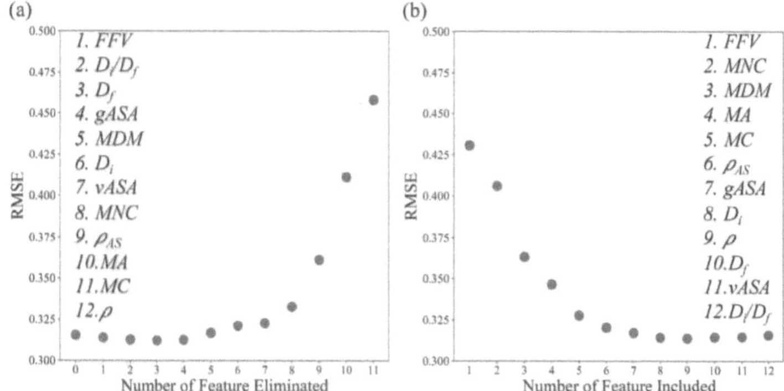

Figure 3.17 Wrapper feature selection method of (a) backward elimination and (b) forward selection. The order of feature eliminated and feature included are specified in the figure.

To avoid the greedy methods of the forward selection and backward elimination, we conduct an exhaustive search of all the possible combinations for a given number of features, which is done for varying number of features. The RMSE with different number of features is shown in Figure 3.18, where the corresponding combinations of the features are also shown. The newly designed features, especially the MA and MC, are included in almost all cases except for when one or two features are allowed for the modelling. When only one or two features are used in the modelling, the performance is very poor as indicated by the RMSE value of over 0.40. This value is very similar to when using only the standard geometrical and chemical features (i.e., without the newly designed features, see Figure 3.14 (a) and (b)). Similar to the forward selection and backward elimination, when the features that relates to the open-metal sites are included, such as when using 3 or

4 features in the exhaustive search, the performance of the model improved dramatically (i.e., fast decreasing RMSE). However, when considering more than 5 features, the RMSE barely decreases. The search again indicates that the newly designed features indeed contain useful information.

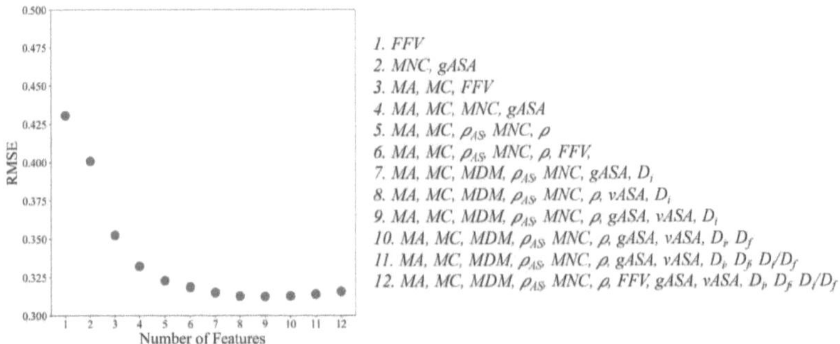

Figure 3.18 Exhaustive search of the features for training the random forest regressor.

3.3.4 Insights on useful features for designing optimal materials

The feature selection procedures discussed above have clearly illustrated that several features, particularly those newly designed ones in this work, are significant predictors of the selectivity. For this reason, a more thorough analysis on these features, (i.e., MA, MC, and ρ_{AS}) is conducted for those highly selective structures. Figure 3.19 shows plots of the correlations between MA, MC, and ρ_{AS}, with highlights on top 20, top 50, top 100, and top

200 highly selective structures. In line with above, a commonality among many of the top 20 structures, top 50 structures, and top 100 structures (Figure 3.19 (a), (b), and (c), respectively) is that these structures possess metals that are geometrically accessible to adsorbates (i.e., $MA > 0$ °) together with strong charge (i.e., $MC > 1.0$ e) that can offer strong interaction with the H_2S. Especially, it can be observed that there are a few structures that have MA of 65° with $MC > 0.7$ e included in the highly selective structures. MA of 65° indicates that these structures have metal sites with only two linkers attached. Furthermore, only a very small portion of these structures have MA below -10°. This trend can also be observed from the distribution of the MA and MC for those highly selective structures as shown in Figure 3.19 (e) and (f). The distribution of MA and MC for the highly selective structures is different from that of the total set of structures. Specifically, highly selective structures tend to have relatively higher population for those higher MA and MC as compared to that of the total set of structures. For the top 100 or top 200 structures (Figure 3.19 (c) and (d), respectively), structures that do not possess open-metal sites but still show high performance start to appear. As expected, these structures tend to possess a high surface atom density (i.e., ρ_{AS}), as a high ρ_{AS} can also provide sites that can effectively interact with adsorbates.

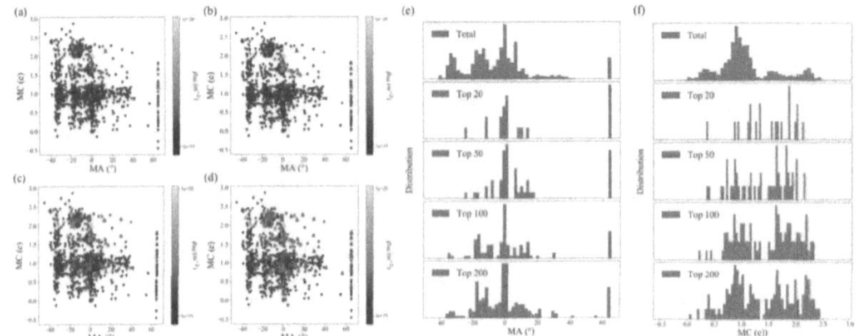

Figure 3.19 Metal charge (MC) vs. metal angle (MA), colored by surface atom density (ρ_{AS}) for all the data sets considered in this study with (a) top 20, (b) top 50, (c) top 100, and (d) top 200 highly selective structures highlighted by triangles. Distributions of MA and MC for those selected structures are shown in (e) and (f), respectively.

3.4 Conclusion

In our study, we employed machine learning algorithms, specifically random forest regressors, to train the selectivity of H_2S/CH_4 for sour gas sweetening with geometrical and chemical features of MOFs. These included the standard geometrical and chemical features established previously, such as crystal density, fractional free volume, surface area, pore diameter, and most negative charge. In addition, we designed new features that are highly predictive of the selectivity. This study demonstrated that those commonly adopted geometrical and chemical features together with our newly designed features that contain important information can yield a good prediction of the H_2S/CH_4 selectivity; the machine learning performance was improved greatly by incorporating these newly developed features. Among these features, fractional free volume and the newly designed features including the metal angle, metal charge, metal dipole moment and surface atom density were considered to be most powerful predictors of the selectivity. The inclusion of other features, such as the pore diameters or accessible surface area, barely affected the performance of the machine learning model. This could be attributed to the high correlations between the standard geometrical features. Analysis on the significant features, specifically the newly designed features including metal angle, metal charge, and surface atom density, informs that it is advantageous to design new materials to have open-metal sites with strong charges. Furthermore, designing materials to possess high surface atom density is beneficial for high selectivity of H_2S. On the other hand, structures that do not possess open-metal sites and at the same time have low surface atom density may unable to offer high H_2S/CH_4 selectivity for effective separation. The selected features can be employed for modelling with other machine learning algorithms that may suffer from overfitting due to irrelevant features for further studies. The features designed herein can also be useful for other applications, especially for those molecules that can benefit from the strong electrostatic field in the open-metal sites, such as carbon dioxide. Overall, our work paves the way for future efficient computational discovery and design of nanoporous materials.

Chapter 4. Nanoporous Material Recognition via 3D Convolutional Neural Networks: Prediction of Adsorption Properties

4.1 Introduction

Nanoporous materials, such as zeolites and metal-organic frameworks, have been investigated for various energy and environmental-related applications including methane adsorption, carbon capture, or sour gas sweetening.[2,8,10,112,129] However, the total number of nanoporous materials can be as much as hundreds of thousands, which makes materials search extremely challenging purely via trial-and-error or brute-force based search.[14,15,17,18] To this end, predictions with machine learning models have recently drawn considerable attention for facilitating the materials discovery.[105,130,131] For these studies, a few geometrical features of the materials are computed and are used to train machine learning models to predict adsorption properties.[91,103,106,132–138] These features, such as crystal density, fractional free volume, surface area, and pore diameters, represent the structure geometry, have clear physical meaning, and are relevant to the adsorbate-adsorbent interaction and adsorption properties.[28,59,117,139–145] Although some useful predictors have made the predictions decently successful, however, they may impose human bias and require human expertise on the exploration and selection of the features. Furthermore, these studies require multiple steps of preprocessing, feature construction, prediction, and feature selection, requiring human labor to proceed step-by-step for modelling.

One could instead let the machines learn the features by itself. One very applicable algorithm for such purpose is convolutional neural network (CNN). CNN hailed from a study of brain's visual cortex in the 1950s by Hubel and Wiesel who found that some neurons of our visual cortex react only to low-level features, such as vertical, horizontal, and other simple lines, which are combined to observe more complex higher-level features.[146,147] CNN has been widely applied to visual perception such as image search services or autonomous vehicles, where the machines read images and self-learn patterns of objects.[148–151] In a similar vein, CNN can be applied to read and detect materials, where the machines will detect low-level lines and curvatures of pores and channels, which then can be combined to higher-level geometry of the structure. Overall, by "looking" at the structures, the machines will be able to learn important features of the structure by itself (i.e., feature recognition in convolutional layers) and predict its adsorption properties (i.e., prediction in fully connected layers). As an illustration, Figure 4.1 shows an example of how our designed CNN model filters through a given structure and makes predictions on its adsorption isotherm. Furthermore, the training and prediction with CNN can be conducted in an end-to-end approach, where the structure is directly employed as an input and the predictions carried out in one step.

In this work, we present the application of CNN for predicting adsorption properties of nanoporous materials, with methane adsorption in zeolites as a case study. Due to the environmental burdens brought about by fossil fuels, natural gas or landfill gas can be great alternatives.[2] However, these energy sources may be contaminated with acidic gases.[31,91,102,152] Furthermore, methane has very low volumetric energy density, as low as one-third of gasoline, which makes transportation and storage of gases extremely difficult.[153] For this purpose, nanoporous materials can be beneficial for methane separation and storage. For instance, Kim et al. have validated that zeolite structures can be effective adsorbents for methane uptake due to the non-polar nature of the molecules.[95]

4.2 Methodology and Computational Details

4.2.1 Data Set

This study utilizes structures from the predicted crystallography open database (PCOD) of Deem's hypothetical zeolite database to train/validate our model.[13] Approximately 6500 structures are selected from the database and these structures have unit cell size below 24 Å in each direction and do not possess inaccessible pockets. We employed Zeo++ with a probe radius of as large as 1.9 Å to identify inaccessible regions.[154] Furthermore, MFI, a widely studied, synthesized structure available from International Zeolite Association[11], is adopted in the study for validation and analysis. For each structure, a cubic cell with size of 24Å in each direction is created by duplicating unit cells and cutting a portion of the super cell created. The structure is then represented with $24 \times 24 \times 24$ voxels where each voxel represents the "availability" of potential adsorption sites in the structures. Namely, the vacancy of framework atoms is marked in each voxel ("Grid Representation" in Figure 4.1, white area represents the vacancy of framework atoms). This is computed with the criteria using van der Waals radii adopted from Mantin et al. (i.e., Si: 2.10 Å , O: 1.52 Å)[155]

The dataset is split into 90 % of the training (i.e., 5830) and 10 % of the validation sets (i.e., 648). Furthermore, in order to ensure that the model be generalized to unseen dataset, data augmentation is applied to the training sets. Although the pooling methods may help the model immune to local translation of the objects to be detected (i.e., adsorption sites), it could still suffer from large translations and rotations of the objects (i.e., model incapable of identifying objects that is translated to a large extent or rotated).[156–158] Thus, the original voxel is translated +12 Å along each direction, with rotations conducted for each created 3D grid, yielding a total of 8×3=24 duplicates for each structure.

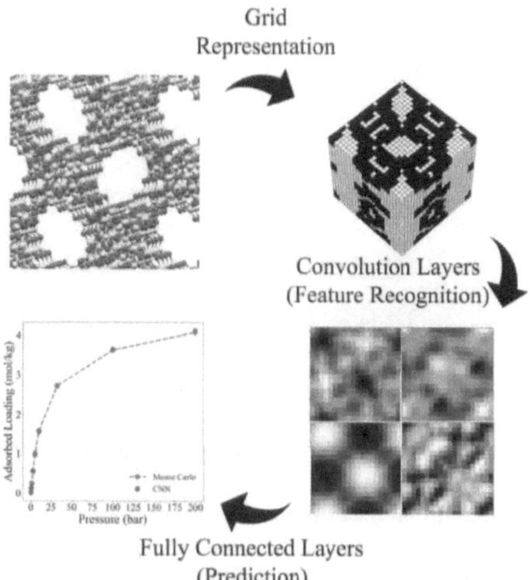

Figure 4.1 Schematics of adsorption isotherm prediction by CNN for structure 8317707 as an example.

4.2.2 Architecture of Convolutional Neural Network

Overall, the architecture of this study resembles that of the LeNet-5 proposed by LeCun et al., most widely known CNN architecture successfully implemented for reading handwritten digits.[159] The LeNet-5 consists of 3 convolutional layers followed by 2 fully connected layers. For the first two convolutional layers of LeNet-5, each convolutions are followed by activation with hyperbolic tangent function (i.e., tanh), average pooling, and then activation with tanh, sequentially. For the last convolutional layer, the convolution is followed only by activation with tanh. The number of kernels (or filters) for each convolutions are 6, 16, and 120, respectively, where the kernel size is 5 with stride of 1 for all cases. For the average pooling, the kernel size is 2 with stride of 2. LeNet-5 does not employ any paddings throughout, except for the zero paddings of input.

This study adopts the number of layers, the number and size of kernels for each layer (except for the last convolution layer in order to compensate for input size of 24 vs. 32), and the number of strides in each convolutional and pooling layer and the number of units in the hidden fully connected layers from that of the LeNet-5 (see Table 4.1). The model is then modified with some more advanced and robust tricks for higher accuracy and efficiency. In detail, batch normalization is used after convolution and before activation. Without the batch normalization, the network may suffer from internal covariate shift; the distribution of the inputs to each layers deep in the network may shift when the weights are updated. Therefore, batch normalization can stabilize the learning process, reduce the number of training epochs significantly, and provide regularization effects (i.e., prevent overfitting and reduce generalization error).[160,161] Additionally, this study uses max pooling in lieu of the average pooling. Pooling has the effect of subsampling the image and thereby reducing computations, memory loads, and number of parameters (i.e., prevent overfitting).[160] Most of all, pooling enables the network immune to a small shift of the structure (i.e., local translation invariance that makes neural network robust to a small shift in the unit cell generated for a given structure). Max pooling may be more suitable in detecting the pore geometry as the average pooling may dilute the contrast in the feature maps.[160,162,163] Furthermore, rectified linear unit (ReLU) is used as activation function as it does not suffer from saturation at positive values and is computationally faster.[164] He uniform initializer, which is compatible with ReLU, is employed for the random initialization of the parameters.[165] For the optimizer, Adam with learning rate of 0.001, β_1 of 0.9, β_2 of 0.999, and epsilon of 10^{-7} is used.[166] These details were modified from the LeNet-5 as they have been employed in recent CNN architectures, such as AlexNet, GoogLeNet, or ResNet, the winners of ImageNet ILSVRC challenge.[148,149,151] Overall, the CNN architecture possesses a total of 76,370 trainable parameters. For optimization, mean squared error (MSE) of logarithm-scaled adsorption loading is used as the loss function. For training, 70-epoch with mini-batch size of 32 is conducted. In order to prevent the model from overfitting, early stoppings are enabled when no further improvements are made in the loss function of validation sets for 30 epochs.

Table 4.1 CNN architecture adopted in this study for predicting methane sorption in zeolite structures.

Layer	Type	Maps	Size of Output	Kernel Size	Stride	Activation
Out	Fully Connected	-	14	-	-	-
F1	Fully Connected	-	84	-	-	ReLU
A3	Activation	-	$1\times1\times1\times120$	-	-	ReLU
B3	Batch Normalization	-	$1\times1\times1\times120$		-	-
C3	Convolution	120	$1\times1\times1\times120$	$3\times3\times3$	1	-
S2	Max Pooling	-	$3\times3\times3\times16$	$2\times2\times2$	2	-
A2	Activation	-	$6\times6\times6\times16$	-	-	ReLU
B2	Batch Normalization	-	$6\times6\times6\times16$	-	-	-
C2	Convolution	16	$6\times6\times6\times16$	$5\times5\times5$	1	-
S1	Max Pooling	-	$10\times10\times10\times6$	$2\times2\times2$	2	-
A1	Activation	-	$20\times20\times20\times6$	-	-	ReLU
B1	Batch Normalization	-	$20\times20\times20\times6$	-	-	-
C1	Convolution	6	$20\times20\times20\times6$	$5\times5\times5$	1	-
In	Input	1	$24\times24\times24$	-	-	-

4.2.3 Monte Carlo Simulation

Each structure is labeled with pure methane adsorption isotherm at a temperature of 300 K with a wide range of pressure points (0.0005, 0.001, 0.005, 0.01, 0.05, 0.1, 0.5, 1, 2.5, 5, 10, 32.09, 100, and 200 bar). Adsorption isotherm of CH_4 in zeolite structures, are computed with Monte Carlo simulations in a grand canonical ensemble (GCMC). In these calculations, the interaction energy is modeled with Lennard-Jones (L-J) potentials with parameters adopted from García-Pérez et al.[167] The L-J potentials are truncated at a cutoff of 12 Å and shifted to zero beyond the cutoff. Periodic boundary conditions are implemented, where each of the simulation cell has a perpendicular cell width at least twice the cutoff (i.e., 24 Å) along each direction. Monte Carlo simulation for MFI is conducted herein using the RASPA package[44], whereas the calculations for PCOD structures are adopted from a prior study.[95]

4.3 Results and Discussion

As can be seen in Figure 4.2, CNN accurately predicts the adsorption loading, shown in both logarithm and original scales. Plot of loss functions as a function of epoch number in Figure 4.3 (a) also shows that the loss function decreases as the epochs increase, with only minor improvements at large epochs. To be specific, the MSE can be as low as 0.0031 and 0.0047 for the training and validation sets, respectively, indicating a good fitness and generalization. Particularly, the low generalization error can be attributed to the data augmentation of training sets. As can be seen in Figure 4.3 (b) and (c), with the absence of data augmentation, increasing the number of structures in the training set after a certain amount (e.g., > 70 %) would not be able to reduce the generalization error. Even with full datasets as shown in Figure 4.3 (a), the gap between the training and validation sets indicates that the model could be suffering from a slight overfitting, but this gap is greatly reduced by introducing data augmentation (see also Figure 4.4). Our results show that, although CNN for computer vision applications may require rich data regimes, its application to materials science could work well with a rather small dataset size. A dataset of 4000 (~70%) is found to be sufficiently large to develop a CNN model to predict methane adsorption in zeolites, and data augmentation represents a very effective way of improving the accuracy of the model. Enlarging the dataset size is also not desirable, as generating dataset is typically the most computationally expensive part of the process. However, we note that it may require a larger dataset when building CNN models for more complex molecules, such as CO_2 that is a non-spherical, linear molecule with a strong quadrupole moment.

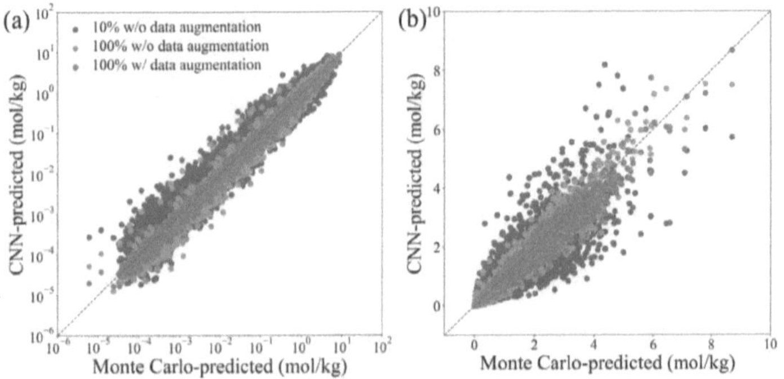

Figure 4.2 Parity plot of CNN-predicted methane loading vs. Monte Carlo-predicted (ground truth) methane loading in (a) logarithm scale and (b) original scale.

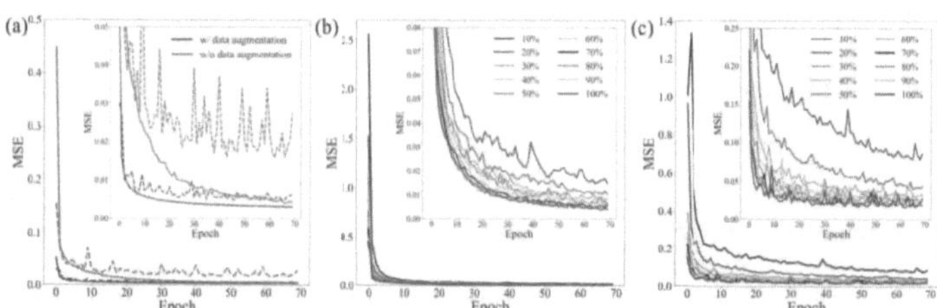

Figure 4.3 (a) Loss function (i.e., MSE) as a function of epoch number trained with full dataset of training set (5830 structure) with and without data augmentation, where solid lines indicate training sets and dashed line validation sets. Loss function as a function of epoch number for (b) training sets and (c) validation sets trained with different portion of the full dataset of the training set (without data augmentation).

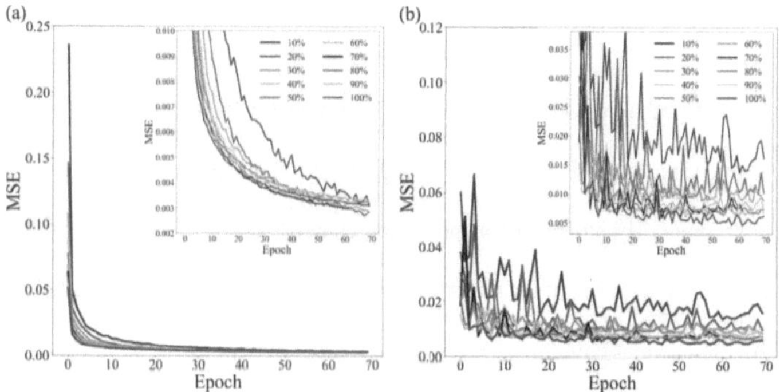

Figure 4.4 Loss function (i.e., MSE) as a function of epoch number for (a) training sets and (b) validation sets trained with different portion of the full dataset of the training set (with data augmentation).

As mentioned previously, a very important advantage of CNN is that the kernels (or filters) can self-learn important features of zeolite structures that most contribute to the adsorption isotherm. The six kernels of the first convolutional layer, which can detect low-level features, are shown in Figure 4.5, with a two-dimensional representation extracted from the three-dimensional kernels. For a better illustration, the kernels of the same layer trained with only 10 % or full dataset of the training set structures without data augmentation are also shown in Figure 4.5 for comparison. For the kernels trained with only 10 % of the training set without data augmentation, it does not show any specific patterns that could be easily interpreted, which may be related to its relatively lower accuracy in the predictions with this CNN (see Figure 4.2 for its predictions). With 10-fold higher training sets (middle figure of Figure 4.5) that can offer much better predictions, some kernels can be interpreted as representing pore-like features, as highlighted by the red circles. After applying the data augmentation (most right figure of Figure 4.5), features that illustrate patterns that can be more obviously interpreted as pores and channels emerge. To be specific, the blue-circled one resembles straight channels whereas the red-circled ones pore-like features. Although the kernels that could not be easily interpreted may possibly be due to the complexity of the structure geometry and two-dimensional representation of the three-dimensional kernels (see Figure 4.6 for other views of the kernels), the number of kernels that obviously delineate pores and channels increase with more training sets and data augmentation (i.e., higher accuracy, which is in line with the high accuracy in isotherm predictions as shown in Figure 4.2). This illustrates that CNN self-learns geometrical features of the structures that resemble channels and pores for providing accurate predictions.

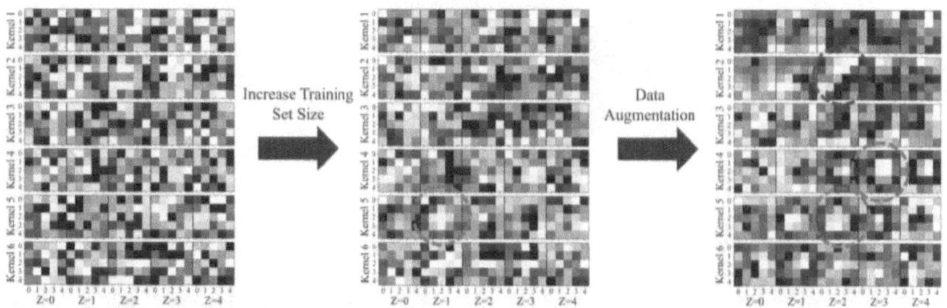

Figure 4.5 Two-dimensional representation (x-y plane for varying z) of kernels of the first convolutional layer. The left figure shows the kernels trained with 10% of the training set (583 structures) and the middle one with full dataset of the training set (5830 structures), both without data augmentation. The most right figure is kernels trained with full dataset of the training set (5830 structures) with data augmentation. The weights are normalized to 0 to 1, where darker color indicates closer to 1.

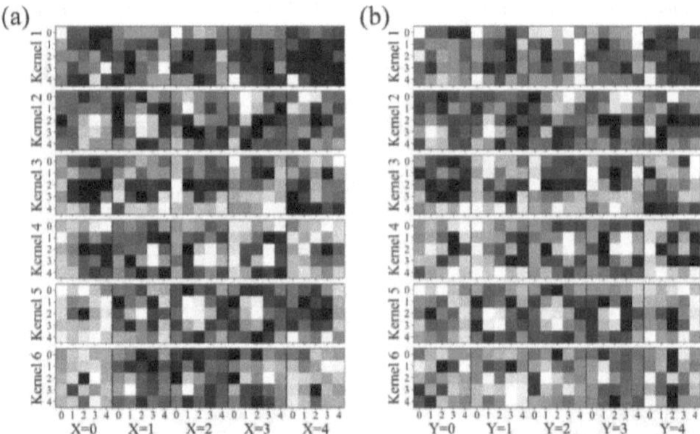

Figure 4.6 Two-dimensional representation, with (a) y-z plane for varying x and (b) x-z plane for varying y of kernels of the first convolutional layer trained full dataset of the training set (5830 structures) with data augmentation. The weights are normalized to 0 to 1, where darker color indicates closer to 1.

We also investigate how the input structures are convoluted with the self-learned kernels to the first hidden neurons of CNN. Figure 4.7 shows an example for a structure that is shown to possess the highest loading in intermediate pressure range among the validation sets. This structure is a one-dimensional structure, where pores are separated with smaller windows. A two-dimensional representation of this structure can be seen in Figure 4.7 (a) and the corresponding grid in Figure 4.7 (b). Figure 4.7 (c) shows the convoluted neurons in two-dimensional representation. It evidently shows that, with more training sets and data augmentation that resulted in higher accuracy in the predictions, the original structure is more clearly detected by the kernels (i.e., clear and well-defined). This again indicates that the kernels are trained to best read the input structure. Specifically, as was expected from Figure 4.5 (red circles that characterize pores), Kernel 4 and Kernel 5 activated the circular shapes of this particular structure. Figure 4.8 exhibits another example for a structure that has the highest adsorption capacity in high pressure range. This structure has more complicated three-dimensional porous geometry, with rectangular-shaped channels along the z-axis. For this structure, the output of convolution operations demonstrates that different kernels played distinct roles; Kernel 2 or Kernel 6 identifying rectangular shapes of channels and framework atoms, whereas Kernel 5 exactly locating the pores. Each of these kernels recognizes different low-level patterns of local receptive fields for a given structure, which are propagated and combined to observe the global structural information that is relevant to the adsorption isotherm.

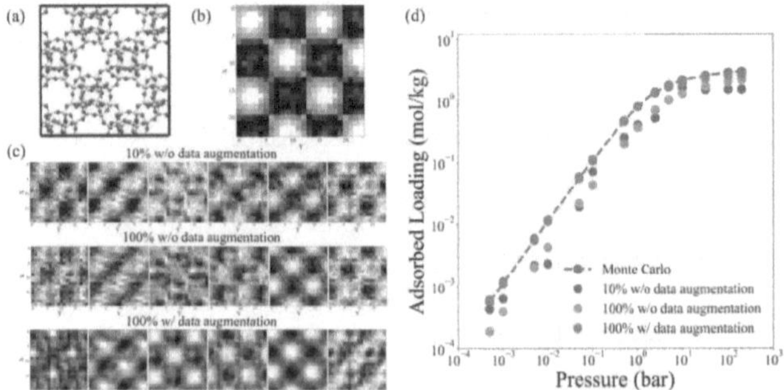

Figure 4.7 (a) Two-dimensional representation (x-y plane) of structure and (b) the corresponding grid for structure 8302007. For (b), the original 24×24×24 grid is summed along the z-axis (direction along the channel) to make 24×24 grid. (c) Output of first convolutional layer (i.e., C1 of Table 4.1) and (d) predictions with CNN trained by different training set size and with or without data augmentation. For (c), six of 20×20×20 grids are extracted from the original 20×20×20×6 grid and are summed along the z-axis (direction along the channel) to make six of the 20×20 grid.

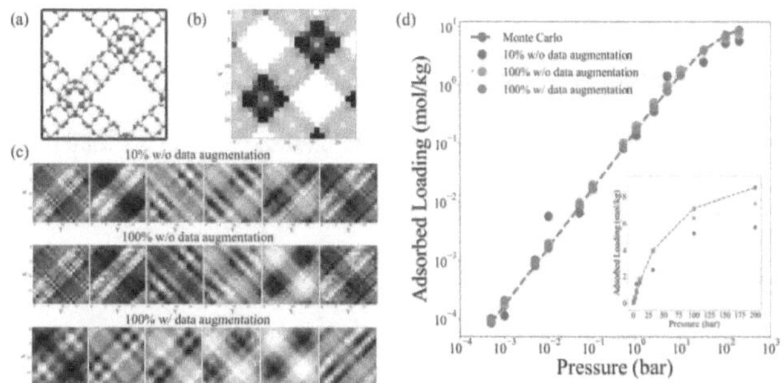

Figure 4.8 (a) Two-dimensional representation (x-y plane) of structure and (b) the corresponding grid for structure 8318323. For (b), the original 24×24×24 grid is summed along the z-axis (direction along the channel) to make 24×24 grid. (c) Output of first convolutional layer (i.e., C1 of Table 4.1) and (d) predictions with CNN trained by different training set size and with or without data augmentation. For (c), six of 20×20×20 grids are extracted from the original 20×20×20×6 grid and are summed along the z-axis (direction along the channel) to make six of the 20×20 grid.

The ultimate of goal of this study is to utilize CNN to facilitate the materials discovery. We show the applicability of CNN to MFI, a widely studied zeolite available from International Zeolite Association that is not included in the training set, and compare with the conventional Monte Carlo simulation methods.[11] A comparison of Monte Carlo-predicted and CNN-predicted adsorption isotherm is shown in Figure 4.9. The prediction is highly accurate, with MSE being only 0.0061 and 0.015 mol/kg in logarithmic and original scale, respectively. This difference between the CNN-predicted isotherm and the GCMC-predicted one is nearly negligible. Regarding the efficiency, calculation of the adsorption isotherm with Monte Carlo simulation consumed a total of more than 56 CPU-hours, whereas less than 0.02 seconds (excluding the grid generation) with CNN. This orders of magnitude difference in the resources needed for the calculations may become tremendous given that the number of nanoporous materials expands as much as hundreds of thousands.

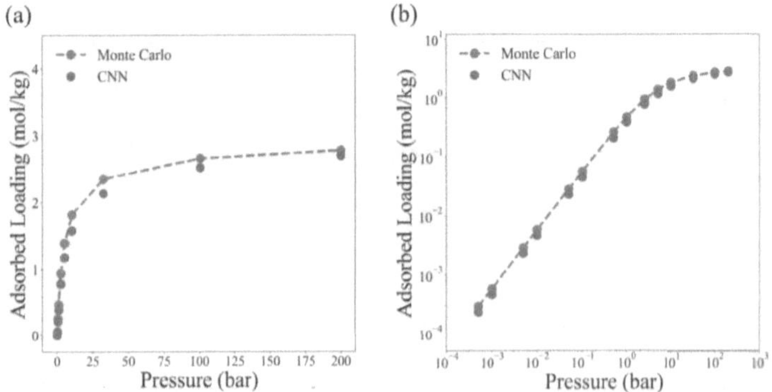

Figure 4.9 Adsorption isotherm of methane in MFI computed with Monte Carlo and CNN in (a) original and (b) log scale.

It is important to stress at this point that the CNN approach introduces a new way of making efficient predictions of adsorption properties for nanoporous materials. Previously, geometrical information, ranging from simple properties such as pore diameter, surface area, void fraction, and others, to rather complicated mathematical techniques to encode them, such as persistent homology, have been designed and constructed by human expertise to serve as features in machine learning algorithms.[91,137,139,168–170] In the CNN approach, the structure is directly employed as the input and the CNN self-learns useful geometrical information most relevant to adsorption properties.

4.4 Conclusion

In summary, we showed in this study the usefulness of 3D convolutional neural network (CNN) in nanoporous materials recognition for predicting their adsorption properties. Just as the 2D CNN has successfully been applied to computer vision applications, this study shows the applicability of 3D CNN in materials science and chemistry for reading the geometrical information of nanoporous structures and making accurate predictions. The success in the predictions can be ascribed to these factors including but not limited to: the power of CNN that self-learns useful features that are common to different local receptive fields of a given structure and can help to avert from human-biased feature construction, the architecture of CNN which already have performed well for reading handwritten digits, well-established and sufficient number of training sets, and regularization effects thanks to data augmentation. We anticipate that CNN can potentially be utilized to other applications in materials science, for instance carbon capture or sour gas sweetening with a variety of nanoporous materials including zeolites, metal-organic frameworks, or porous polymer networks.

Chapter 5. Concluding Remarks and Future Work

Computational modelling methods, together with machine learning algorithms, are employed in order to facilitate the search of nanoporous materials, such as zeolites and metal-organic frameworks (MOFs), for energy and environmental applications. Specifically, in our study, Monte Carlo simulations are implemented for computing adsorption properties and the generated data are utilized to train machine learning models for more efficient predictions. In order to ensure the success of machine learning predictions, it is critical that the data used for training and validations are reliable. For such purposes, we firstly focused on developing accurate molecular models, which would be critically important for the accuracy of Monte Carlo simulations. We proposed a new scheme for molecular model development, where the relevant model parameters are firstly optimized to reproduce electrostatic potential surrounding the molecule and then remnant parameters are fitted to experimental vapor-liquid equilibrium. Our methodology enabled accuracy of the models in reproducing the electrostatic potential surrounding them and efficiency in the model development by decoupling the model parameters. This methodology are applied to various gas molecules related to energy and environmental issues, including CO, CO_2, COS, H_2S, N_2, N_2O, and SO_2. Our analysis and results show that the developed model proves to be highly accurate in reproducing binding mechanism and adsorption properties. The accurate molecular models are then utilized in Monte Carlo simulations to compute their adsorption properties in nanoporous materials, which are then employed to train machine learning algorithms. For reliable machine learning predictions, it is also important to have descriptors that can sufficiently and adequately represent the target value of interest. Our study approaches in two directions for designing descriptors: human-designed descriptors by taking advantage of human knowledge and expertise and machine-learned descriptors where machines self-educate useful descriptors for the predictions. Considering that open-metal sites are strong binding sites for molecules with strong dipole or quadrupole moments, we designed new descriptors that could quantitatively represent the "openness" of these sites. By incorporating them in machine learning models, together with other conventional descriptors of MOFs, the predictions are improved and decently accurate. On the other hand, we made use of convolutional neural networks (CNNs), which have been widely applied in computer vision applications, for recognizing nanoporous materials, with self-learned descriptors, and making predictions of adsorption properties. This is implemented in an end-to-end manner, where the structure is directly used as input and the CNN makes predictions by reading the structure. Our results show that the CNN can make highly accurate predictions and applicability of CNN in materials science.

Future work can be done by utilizing our developed model and methodology for facilitating materials discovery for various applications. Our developed ESP-MMs include various molecules that are environmentally important. However, other than CO_2 or N_2, their study have not yet been conducted extensively so far. Our methodology for model development can also be further extended to other molecules. For instance, more molecular models can be developed, such as paraffins and olefins, which are larger and complicated than ESP-MMs. These models can also be a decent cornerstone for optimizing adsorbent-adsorbate interaction parameters. While the molecular simulation can be computed with our developed molecular models, their calculations can be predicted more efficiently with the use of machine learning algorithms. This can be done by machine learning algorithms and/or descriptors that we developed in our study, such as random forest regressor, convolutional neural network, and descriptors that portray open-metal sites of metal-organic frameworks. Particularly, the descriptors regarding the open-metal sites would be beneficial for system that contains molecules with strong dipole or quadrupole moments. In addition, convolutional neural networks can also be employed for accurate and efficient predictions in a variety of systems. This can be done with transfer learning of our optimized model that can enable efficient learning of the parameters for same materials but with different gas adsorption, such as CO_2 adsorption in zeolites. Additionally, machine learning algorithms other than we have explored herein can be examined to find ways to facilitate the materials search for various applications.